Public Health Risks of the Dioxins

Public Health Risks of the Dioxins

Proceedings of a Symposium held in New York City on October 19-20, 1983
by the Life Sciences and Public Policy Program
of The Rockefeller University

William W. Lowrance, *Editor*

Library of Congress Cataloging in Publication Data

Main entry under title:

Public health risks of the dioxins.

Bibliography: p.
Includes index.

1. Dioxins—Toxicology—Congresses.
2. Tetrachlorodibenzodioxin—Toxicology—Congresses.
3. Environmentally induced diseases—Congresses.
I. Lowrance, William W., 1943-
II. Rockefeller University. Life Sciences and Public Policy Program.
RA1242.D55P83 1984 615.9'513 84-17987
ISBN 0-86576-076-4 AACR2

Distributed by William Kaufmann, Inc. 95 First Street, Los Altos, California 94022

Printed in the United States of America

Contents

Preface and Acknowledgments

On October 19-20, 1983, the Life Sciences and Public Policy Program of The Rockefeller University convened a symposium on the Public Health Risks of the Dioxins. This conference was meant to follow upon a previous symposium, held in June 1981, on Assessment of Health Effects at Chemical Disposal Sites.*

Three other institutions served as co-conveners: the Sloan-Kettering Institute of the Memorial Sloan-Kettering Cancer Center (Richard A. Rifkind, Director); the Environmental Sciences Laboratory of the Mt. Sinai School of Medicine (Irving J. Selikoff, Director); and the Division of Environmental Sciences of the Columbia University College of Physicians and Surgeons (I. Bernard Weinstein, Director).

Participants included a wide variety of scientific experts and leaders from industry, government, special-interest groups, academia, and the press (listed at end of this volume). Selection of the 400-some participants was based on advice gathered by iterative polling of the leaders of organizations that have worked on the dioxin problem. William W. Lowrance, Director of the Life Sciences and Public Policy Program, chaired the symposium.

Experts in specialties from analytic chemistry to pathology to biostatistics, approached on short notice, delivered presentations in the meeting. Lively panel and audience discussions followed. After the symposium the authors, joined by colleagues, prepared the papers published here. Thus these articles reflect the discussions during the meeting, and afterthoughts.

This symposium was organized as a meeting of scientists and physicians to discuss the difficult technical issues, not to argue about political, legal, economic, and other nonscientific issues, which need different forums. Good-faith respect for these intentions was observed by the participants during the entire two-day symposium.

As became exhaustingly clear during our fundraising efforts, three kinds of projects are anathema to most philanthropic foundations: conferences, conferences on controversial subjects, and conferences leading to publications. This project, attempting to meet a pressing public need, was of course all three. And our leadtime was much shorter than foundations usually require. Therefore we are extremely grateful to those who responded to our urgent request for funding:

Charles Stewart Mott Foundation
Merck & Co., Inc.
Rockefeller Family Fund
Syntex Corporation
The New York Times Company Foundation.

Because no honoraria and few travel funds were available, it is a tribute to the dedication of the 400 participants and their institutions that almost all of them paid their own expenses to take part in this conference.

Susan Sheridan administered all the preparations for the symposium. The staff of Automated Text, Inc., stalwartly "processed" the drafts of the manuscripts through many revisions. And Carol Moberg served resourcefully as managing editor to direct the publication of these proceedings.

*Assessment of Health Effects at Chemical Disposal Sites (William Kaufmann, Inc., Los Altos, CA, 1981).

Interpretive Summary of the Symposium

William W. Lowrance

(This summary and commentary on the symposium seeks to describe the background on dioxins, summarize the principal themes of concern and areas of apparent agreement, point out areas of uncertainty and disagreement, list some ongoing activities, and suggest implications for research and policymaking. This is not a consensus document, but an interpretation.)

PURPOSE OF THE SYMPOSIUM

The symposium was convened to: (a) develop critical review of scientific issues surrounding the human health risks from low-level exposure to the dioxins; (b) address scientific questions relevant to impending public policy and managerial decisions, and to the research agenda; (c) examine the dioxins as prototypes of other issues of this kind that will be arising; and (d) offer the newsmedia an opportunity to summarize and interpret these issues for the public.

The symposium did not discuss details of the origins of the dioxins or approaches to environmental decontamination, but focussed on biological and health issues.

The goal was to encapsulate, critique, and discuss current understanding, draw out lessons for other such hazard situations, and make suggestions for action.

OVERVIEW OF THE DIOXINS PROBLEM

The dioxins are a family of 75 closely related compounds that occur as trace environmental contaminants. They constitute a major public concern and toxicological puzzle. Their risks to humans still are not well understood or precisely estimated.

Occurrence of the dioxins. The dioxins have never been manufactured deliberately, although small amounts have been synthesized in laboratories for experimental use. In some situations these compounds have been generated as inadvertent and unwanted by-products in manufacture of chemicals based on chlorinated phenols. In some cases, such as in phenolic wood-preservation uses, and in areas sprayed with the herbicide "2,4,5-T" [(2,4,5-trichlorophenoxy)acetic acid, containing residual dioxin contaminants] for civilian forestry or weed-control or military defoliant programs, the dioxins are dispersed and are slowly decaying. In other cases, such as at industrial sites where these materials have been handled, small concentrated pockets of the toxins exist. In a few cases, such as at the former hexachlorophene plant at Seveso, Italy, or on roadways and farms in Missouri, the materials have been released by unusual circumstances -- at Seveso by explosion of the plant, at Times Beach by spraying of waste oil (containing dioxins) for dust control.

There have been some occupational exposures, in manufacturing, in formulation and spraying, and in cleanup of spills and other releases. Fires involving heavy electrical equipment can release dioxins, that either exist in polychlorinated biphenyl (PCB) electrical insulating fluid, or are formed from the PCBs or contaminating chlorobenzenes by oxidation in the fire. Also, it is becoming clear that under some circumstances when hydrocarbons are incinerated in the presence of chlorine radicals, traces of dioxins can be produced in the flames; the extent to which this can contribute to the environmental dioxin burden is now being investigated. Different geneses produce different mixtures of the 75 dioxin compounds.

Search and cleanup. At present, extensive tests are being conducted in a number of countries to analyze for dioxins in air, soils, waters, fish, foods, and animal and human tissues.

Because of health concern, extreme precautions now are being taken to prevent formation and release of dioxins in most situations where they could possibly be generated. Where dioxins have been found, cleanups are being undertaken. Research is underway to devise methods for ultimate destruction of the dioxins collected in cleanups.

Physical properties. The dioxins are extremely stable physically and biologically. In the environment they do decay, but slowly. The dioxins are virtually insoluble in water, but are soluble in organic solvents and fats and oils. (Thus they tend to precipitate to the bottoms of waterways, and to concentrate in the fatty tissues of fish, animals, and humans.) Dioxins adsorb onto soil particles and are not easily desorbed. They do not burn readily.

TCDD. For several decades research has focussed primarily on 2,3,7,8-tetrachlorodibenzo-p-dioxin (TCDD), which on most tests appears to be the most toxic dioxin. A principal issue for current research, now that many other isomers besides TCDD are being studied, is to sort out the extent to which TCDD should be considered the bellwether isomer with respect to health effects.

The furan and PCB connections. A flanking issue should be recognized at the outset: several other sets of compounds that closely resemble the dioxins need to be kept in mind. The dibenzofurans (135 compounds), which are structurally similar to the dioxins, are generated as by-products of some minor manufacturing processes, can be released in electrical equipment fires, and occasionally are encountered from other sources. Besides, many samples of PCBs, upon sensitive analysis, are turning out to contain traces of the dibenzofurans; there is serious speculation that these furans may account for much of the toxicity of PCB fluids. A few other related classes of compounds (chlorinated azoxybenzenes, chlorinated naphthalenes, etc.) also are used in research and commerce.

It was the sense of the symposium that because these compounds appear to be at least as toxic to animals as the dioxins, and occur in much the same kinds of situations, they need to be studied along with the dioxins.

Reasons for health concern. Without question, the dioxins are extraordinarily toxic to rodents and other lower animal species, as is clear from many of the papers in this symposium. The effects are very dose-dependent.

However, despite passionate concern on the part of many possibly exposed individuals, and despite extensive scientific investigation, it is not obvious that the compounds are so toxic to humans. From several human occupational and accidental incidents, it is evident that at fairly high levels, dioxin exposure causes chloracne (a severe, persistent skin rash) and a variety of other toxic responses. Although chloracne is a reliable sentinel of exposure, absence of chloracne cannot be taken as proof that exposure has not been incurred. Epidemiologic follow-ups have not yet convincingly revealed any increased or unusual pattern of mortality from human exposure. Several symposium papers pursue the controversy surrounding this issue. (The symposium was not able to marshal and review all the direct evidence on humans. Some has been published recently, as cited at the end of this summary.)

Thus the dioxins pose a classic public health dilemma: they are extremely toxic to test animals, but are not clearly so toxic to humans. Human experience is accumulating only slowly. Moreover, most of the human exposures appear to have been very small, and have been incurred under circumstances, such as accidents, sporadic spraying, and war conditions, that make scientific analysis of the exposures and effects very difficult.

THEMATIC ORGANIZATION OF THE SYMPOSIUM

The symposium discussion proceeded from analytic-chemical issues, to human tissue analyses, to small mammal toxicity, to carcinogenicity in rodents, and then to the issue of the principal conjectured human cancer (soft tissue sarcoma), and on through immunotoxicity, reproductive toxicity, and broad metabolic alterations. Then came overviews of actions currently being taken by federal agencies. This was followed by panel discussions of research strategy and agenda. These proceedings are organized in the same way, except that the panel and audience discussions have been woven into the papers or this summary. This overview can only serve to indicate the rich detail carried in the papers themselves.

THE SYMPOSIUM PRESENTATIONS

Overview of biological effects of TCDD in experimental animals. Robert Neal reviewed the great range of effects dioxin has been found to induce in experimental animals. These effects include, variously, induction of the P-450 enzyme system and related enzymes, lethality, a wasting syndrome, hepatic damage, reproductive and teratogenic effects, carcinogenicity, and other toxicities.

The acute lethality of TCDD poses a puzzle, in that the median lethal dose (LD_{50}) differs widely among rodent species: at the extremes, the LD_{50} for the guinea pig is 2 ug/kg, and that for the hamster is over 3,000 ug/kg. Even different strains of mice vary among themselves in susceptibility.

One biochemical mechanism of effect is widely confirmed, and that is the induction of enzymes mediated by a cell-surface receptor (the *Ah*-locus receptor). But while this accounts for many effects, it may not necessarily account for all the kinds and intensities of effects seen in different experiments. Other mechanisms also are being postulated [see, for instance, the paper by Matsumura et al. in this symposium].

In the animal studies it is not yet settled to what extent biological effects are caused by metabolites of TCDD as opposed to TCDD itself. Dr. Neal concluded that "the acute toxic effects of TCDD appear not to be related, at least directly, to the rate of metabolism in experimental animals nor to the half-life of excretion."

Analytic chemistry and human fat analyses. Analyses have been conducted on blood, milk, and other human fluids and tissues, but usually this has pressed the techniques to the very limits of detection. Dioxins concentrate preferentially in fat tissues and are most readily detected in that material.

Thomas Tiernan described the current capabilities and limitations of methods for analyzing the dioxins and dibenzofurans. Then he reported some observations of dioxins and furans in human tissues from a variety of sources.

Alvin Young and Christoffer Rappe reported analyses of human adipose tissue conducted in their laboratories and elsewhere. As in the Tiernan laboratory, dioxins (and in many cases furans) were reliably detected, isomer-specifically, at levels on the order of 1-10 parts per trillion (ppt). The compounds have been detected in the inhabitants of Seveso, residents of the Canadian and American Great Lakes area, inhabitants of forested areas of Sweden and Canada, Vietnamese and American veterans of the Vietnam War, and firemen and others exposed to the soot and debris from several building fires.

A major drawback is that for most of the tissue analyses reported so far, it has not been possible to know with any certainty the initial exposure experienced by those people.

In view of the slowness and expense of chemical analyses of environmental samples, suggestions were made that inexpensive rapid bioassays be developed for screening.

Many symposium participants now make the working assumption that most members of the general human population carry extremely low background levels of the dioxins and furans (on the order of 1-10 ppt in adipose tissue). Perry Gehring suggested an approach to thinking about this, which he drafted into a brief "think-piece" [included here] after the symposium.

Carcinogenicity. Richard Kociba described his experimental rodent assessment of TCDD carcinogenicity, and the results of other such assays. In a standard experiment, he found that TCDD at 0.07-0.1 ug/kg/day in diet definitely elicited carcinogenic response, in two strains of rats.

Experiments by Alan Poland and others have shown that TCDD acts as a carcinogenic promoter in a special mouse assay in some strains [the Kociba paper summarizes]. Kociba and others have conjectured that, so far, all data suggest that the dioxins' carcinogenesis proceeds through a non-genetic mechanism.

John DiGiovanni [this symposium] reported on a set of

complicating (and possibly opposing) effects, in which the dioxins and other halogenated compounds induce enzyme production that leads to protective, "anticarcinogenic" action. He concluded: "It may, therefore, be very difficult to determine the outcome of exposure to these agents when exposure to other carcinogenic agents occurs at about the same time."

The Kociba tests have been reviewed extensively and have widely been accepted as definitive. The participants in this symposium did not undermine that acceptance, although they raised questions about details. Robert Squire (Johns Hopkins University) has reviewed the pathological analysis from the Kociba experiment. Christopher Portier et al. [this symposium] have conducted an elaborate statistical analysis, based on the Kociba results, to make an estimate of the cancer risk from TCDD exposure. These data have been adopted by the U.S. Centers for Disease Control in establishing a "level for public health concern" [Kimbrough et al., this symposium].

John Van Ryzin reported on the risk projections he, Christopher Portier, and David Hoel have calculated (see Portier et al. paper for mathematical detail). They concluded: "Using linear modelling, the estimated doses which would yield an added risk of one-in-one-million ranged from 38 fg/kg bw/day in female rat livers to 3000 fg/kg bw/day in female mouse subcutaneous tissue." The discussion that followed this paper raised many of the perennial issues concerning such animal tests: how to aggregate various tumors for statistical reckoning; which statistical models to apply; and what these calculations can be taken to imply for human risk.

Renate Kimbrough reported on how the U.S. Centers for Disease Control (CDC) used these carcinogenicity projections to establish guidelines for concern about contaminated residential soil. In essence, Kimbrough et al. have developed a cascade of estimates that proceed from soil contamination levels, to possible exposure from contact and inhalation, to human cancer risk (the latter based on the Portier et al. analyses of the Kociba et al. rat assays). Based on their cautiously weighted sequence of estimations, they concluded that "a soil level of 1 ppb TCDD in residential areas is a reasonable level at which to express concern about health risks."

[Despite strong qualification by the CDC that this "public-health concern level" is a tenuous, extremely conservative estimate meant only as guidance on soil-borne exposures in residential areas of Missouri, in recent months it has been widely adopted by other authorities, without revision, for many other purposes.]

Bernard Weinstein reviewed current speculations on dioxins as carcinogenic promoters, and expressed strong reservations about the advisability, in environmental regulation, of treating "promoters" differently from "initiators", or epigenetic carcinogens differently from those that appear to act genetically. He recommended that until we know more about the biologic mechanisms, we should treat all substances having carcinogenic properties cautiously and similarly -- not treating TCDD, for example, more leniently than known tumor initiators.

<u>The soft tissue sarcoma problem.</u> The principal form of cancer conjectured to have been associated with exposure to dioxins has been soft tissue sarcoma. Human epidemiologic studies are still being performed, and some that have been completed are quite controversial. No one in the symposium suggested other forms of cancer (these would, of course, be picked up in epidemiologic studies such as worker mortality studies).

Steven Hajdu described the complications encountered in identifying soft tissue sarcomas pathologically. This is a very large family of tumors that can affect many different tissues and take a wide variety of morphologic forms. Too, some non-tumorous pathologies may resemble these sarcomas. Dr. Hajdu criticized the possible misclassification of tumors in some published reports that allege connections between sarcomas and dioxin exposure. He also deplored the publication of medical reports that do not include detailed pathologic information but only summary conclusions. For resolution, he urged formation of a national expert committee to review pathological materials in this problematic area.

Marilyn Fingerhut described reviews conducted by the National Institute for Occupational Safety and Health of seven cases reported as soft tissue sarcoma among persons occupationally exposed to dioxin-contaminated pesticides. These cases have been central in the search for human epidemiologic information. A definitive conclusion on the larger issue has not yet been reached. [Fingerhut et al. later in this symposium describes a NIOSH occupational dioxin registry that is intended to develop further information.]

Ralph Cook pointed out three sources of difficulty in confirming a cause--effect relationship between dioxin exposure and soft tissue sarcomas: inadequacies in some Swedish case-control studies; confusions between national mortality classification schemes and other pathologic classifications; and other epidemiologic complications.

Reproductive toxicity. Donald Mattison reviewed experiments that show impairment by TCDD of endocrine and some other reproductive functions in experimental animals, but pointed out that no conclusions can be reached from the human observations made so far.

Wilbur McNulty summarized studies of TCDD on primates and other mammals. At moderate doses (on the order of 1-10 ug/kg bodyweight), TCDD has caused fetal loss or abortion in several mammalian species. Diminution of fertility, and other reproductive effects, have been observed in some species when TCDD was continuously administered in food in the 100-ppt range. McNulty argued that the female reproductive attributes of rhesus macaques and other Old World primates resemble those of humans, and therefore are the test species of choice. He cautioned, however, that studies on primates at doses in the range likely to be experienced by humans are not likely to produce statistically convincing results.

Maureen Hatch reviewed all the human reproductive studies reported on dioxins, and found that many of them are badly flawed in design. No firm conclusion can be reached yet. A major shortcoming has been that in most cases the dioxin exposure level has not been known or inferrable.

Immunotoxicity. Jack Dean summarized evidence that indicates that TCDD can have powerful effect on the immune systems of experimental animals. In most cases these effects have been dose-dependent, and appear to have been associated with the *Ah*-locus receptor. TCDD exposure diminishes resistance to infectious agents and transplanted tumor cells. The several reported immunological studies of humans accidentally exposed to dioxins are, according to Dean, inconclusive.

Broad metabolic toxicity. Three of the symposium papers reviewed wide-scale, complex metabolic alterations that have been observed in experimental animals.

Fumio Matsumura described a series of dramatic modifications of rat hepatic plasma membrane functions (receptor and enzyme activities) that TCDD induces. These findings may have implications for mechanistic understanding.

Richard Peterson and Carl Potter reported experiments on rats to elucidate the mechanism by which TCDD causes reduced food intake and weight loss and leads to the often-observed "wasting syndrome" in animals fed TCDD. One of the intermediate effects is reduction of the levels of circulating thyroid hormones. Another effect is reduction of plasma testosterones. The authors did not speculate on how these effects may be paralleled in humans.

Masanori Kuratsune proposed a way of establishing clinical indicators for exposure to dioxins, PCBs, dibenzofurans, and related compounds. The method would be based on the analysis of patterns of isomers of residual chlorinated hydrocarbons in blood or other tissues. While there was not any consensus on this particular proposal, many in the symposium seemed to believe that some such pattern analysis might prove useful. A major complicating factor is that different isomer-mixes may induce different mixtures of metabolizing enzymes, which in turn would metabolize the different isomers at differing rates.

Current federal agency activities. Donald Barnes outlined the U.S. Environmental Protection Agency's Dioxin Strategy [the

Executive Summary of which is printed in these proceedings]. This comprises a framework for sampling and analyzing for dioxins in the environment, a scheme for setting regulatory priorities, guidelines for cleanup action, and plans for research.

Marilyn Fingerhut described the U.S. National Institute for Occupational Safety and Health's occupational Dioxin Registry, which is compiling demographic and work-history information for all production workers who have synthesized products known to be contaminated with TCDD or the hexachloro-dibenzodioxins. Currently the Registry includes about 6,000 workers. The Registry is being used as the basis for mortality studies and other attempts to correlate exposure with health effects.

Robert Scheuplein presented the U.S. Food and Drug Administration's approach to tolerance-setting for dioxins in foods. Using worst-case estimations, the FDA has established a "level of concern" of 25 ppt for TCDD in Great Lakes fish.

SOME EVENTS AND PUBLICATIONS SINCE THE SYMPOSIUM

On October 31, 1983, the Missouri Dioxin Task Force submitted its final report to Governor Christopher S. Bond. This report recommended that secure central storage be provided for soil gathered from contaminated areas in Missouri, that health studies of potentially exposed citizens be expanded, and that a (specified) strategy be pursued for accomplishing these ends.

On December 6-7, 1983, the Center for Environmental Toxicology of Michigan State University held a conference on Dioxins in the Environment. The proceedings will appear as Michael Kamrin and Paul Rodgers, editors, Dioxins in the Environment (Hemisphere Publishers, New York, in press, 1984).

Christoffer Rappe published an extensive review, "Analysis of polychlorinated dioxins and furans," in Environmental Science and Technology 18, #3, 78A-90A (1984).

Two sets of papers have been published from special sessions of American Chemical Society Annual Meetings. These appeared as G. Choudary, L.H. Keith, and C. Rappe, editors, Chlorinated Dioxins and Dibenzofurans in the Total Environment (Butterworth, Boston. Volume I, 1983; Volume II, 1984).

Raymond R. Suskind and V.S. Hertzberg published their clinical epidemiologic study of 204 industrial workers known to have been exposed in a 2,4,5-T manufacturing accident in 1949. [Ref: Journal of the American Medical Association 251, 2372-2380 (1984)]

The U.S. Air Force published a study by G.D. Lathrop, W.H. Wolfe, R.A. Albanese, and P.M. Moynahan, "Project Ranch Hand II. An epidemiologic investigaton of health effects in Air Force personnel following exposure to herbicides" (U.S. Air Force, Brooks Air Force Base, San Antonio, Texas, 1984).

BIOLOGICAL EFFECTS OF 2,3,7,8-TETRACHLORODIBENZO-*p*-DIOXIN IN EXPERIMENTAL ANIMALS

Robert A. Neal, Ph.D.

Chemical Industry Institute of Toxicology
Research Triangle Park, North Carolina 27709

From Public Health Risks of the Dioxins, proceedings of a symposium held on October 19-20, 1983 at The Rockefeller University, New York City. Edited by William W. Lowrance. Published by William Kaufmann, Los Altos, California.

Certain isomers of the polychlorinated dibenzo-*p*-dioxins and polychlorinated dibenzofurans produce a number of biological effects in experimental animals. The most prominent of these are enzyme induction, lethality, a wasting syndrome, lymphoid involution, hepatic damage (in some species), chloracne (in a few species), hepatic porphyria, gastric lesions and urinary tract hyperplasia (again only in some species), edema (in certain species), hyperlipidemia, reproductive toxicity, teratogenic effects, and increase in tumor incidence in various organs of rats and mice (1). The concentrations of the select chlorinated dibenzo-*p*-dioxins and dibenzofurans which cause these biological effects are almost always many orders of magnitude lower than the concentrations of commercial PCBs required to cause the same or similar biological effects.

This discussion will be largely confined to the mechanisms of toxicity of the polychlorinated dibenzo-*p*-dioxins and more specifically to the specific isomer 2,3,7,8-tetrachlorodibenzo-*p*-dioxin (TCDD). While, there are 75 possible isomers of the chlorinated dibenzo-*p*-dioxins, the isomer most biologically active in experimental animals appears to be TCDD.

TOXICITY OF TCDD

The single dose of TCDD which produces acute lethality in a number of animal species varies quite widely (Table 1).

Table 1. Single dose LD_{50} values for TCDD.

Species	Route	LD_{50} (ug/kg)	Reference
Guinea pig	Oral	2	(18)
Monkey	Oral	50	(18)
Rat:			
Adult male	Intraperitoneal	60	(21)
Weanling male	Intraperitoneal	25	(21)
3-MC pretreated:			
Weanling male	Intraperitoneal	44	(21)
Adult female	Intraperitoneal	25	(21)

Table 1. Continued.

Species	Route	LD_{50} (ug/kg)	Reference
Rabbit	Oral	115	(23)
Rabbit	Skin	275	(23)
Mouse:			
C57BL/6J	Intraperitoneal	132	(22)
DBA/2J	Intraperitoneal	620	(22)
B6D2F1/J	Intraperitoneal	300	(22)
Hamster	Intraperitoneal	>3000	(19)
Hamster	Oral	5051	(20)

Of the animal species so far examined, the guinea pig (2 ug/kg) (2) and the hamster (>3000 ug/kg) (3,4) occupy the extremes. The median lethal dose (LD_{50}) for TCDD to the remainder of the animals which have been examined is between these two extremes. The LD_{50} of TCDD varies by sex in the Sprague-Dawley rat, with the female (25 ug/kg) being more susceptible than the male (60 ug/kg) (5). The LD_{50} also varies with age in the rat, with the weanling male (25 ug/kg) being more susceptible than the adult male (60 ug/kg) (5). There is a variation in LD_{50} by strain in mice with the C57BL/6J mouse (132 ug/kg) being more susceptible than the DBA/2J (620 ug/kg) (6). As noted in Table 1, the LD_{50} for rabbits also varies by route of administration (7).

Of the various biological effects of the chlorinated dibenzo-p-dioxins and dibenzofurans, the mechanism of induction of the activity of enzymes is best understood. Listed in Table 2 are some of these enzymes whose activities are temporarily increased in animals exposed to TCDD and certain other chlorinated dibenzo-p-dioxins, dibenzofurans, and PCB isomers (1).

Table 2. Enzymes induced by TCDD in experimental animals.

Cytochrome P-450	a-Aminolevulinic acid synthetase
UDP-glucuronyltransferase	Glutathione-S-transferase B
DT-diaphorase	t-Aldehyde dehydrogenase
Ornithine decarboxylase	Choline kinase

The increase in the activity of at least some of these enzymes, on administration of TCDD and certain related compounds to experimental animals or incubation of TCDD with certain cells in culture, apparently results from a binding of TCDD to a receptor protein in the cell, and from the translocation of this TCDD-receptor complex into the nucleus (1). In the nucleus this receptor-TCDD complex apparently binds to a regulatory gene which controls the concentration of most if not all of these enzymes in the cell. This regulatory gene is often referred to as the *Ah* locus after Nebert (8). Although a number of workers have contributed to our understanding of this process, the original observations have come from the laboratory of Alan Poland and his colleagues (9).

Poland has carried out a number of studies correlating the binding of various chlorinated dibenzo-*p*-dioxins to this receptor protein with the biological activity of these same compounds (10). One such study was a comparison of the relative affinity of TCDD and other dibenzo-*p*-dioxins for the receptor protein in the liver cytosol of C57 Black 6 mouse with the ability of these compounds to induce the activity of the P-450 enzyme, AHH, in chick embryos. These experiments showed a good correlation between affinity of the compounds for the receptor and their ability to induce AHH in the chick liver.

Poland's study also demonstrated that only those chlorinated dibenzo-*p*-dioxins in which at least three of the four lateral positions (2, 3, 7 or 8) on the dibenzo-*p*-dioxin ring system are occupied with chlorine atoms have an appreciable ability both to bind to the receptor and to induce AHH in chick embryos at the concentrations used in these studies. The acute lethality of these chlorinated dibenzo-*p*-dioxin isomers in a specific species of experimental animal (2) generally follows the same structure-activity relationship shown in this study for receptor binding and induction of chick embryo AHH (1). Additional work in a number of laboratories using different strains of mice in which there are variable levels of the receptor, as detected by incubation of liver cytosol with ^{3}H-TCDD, has found that the level of the receptor or the affinity of the receptor for TCDD also correlates with the ability of TCDD to induce the activity of AHH (1), to cause acute toxicity (11), to bring about thymic involution (12), to produce cleft palate (12) or induce hepatic porphyria (13). However, this correlation between the level of the receptor and various biological effects in the mouse strains does not hold when other species are analyzed.

CONCENTRATION OF TCDD RECEPTORS

A comparison of the concentrations of the TCDD receptor in the rat and various strains of mice has recently been published (14) (Table 3).

Table 3. Concentration (n) and dissociation constants (K_D) of the TCDD receptor in rat and mouse hepatic cytosol.

Species	n (fmol/mg protein)	K_D (nM)
Sprague-Dawley rat (7)	61 ± 5	0.12 ± 0.03
C57BL/6J mice		
May-July (4)	74 ± 10	0.29 ± 0.01
February-April (3)	47 ± 8	0.29 ± 0.03
DBA/2J mice (3)	N.D.	N.D.
B6D2F1/J mice		
February-April (3)	23 ± 2	0.42 ± 0.03

Gasiewicz and Neal (14).

N.D. = not detected.

With the exception of the concentrations of the receptors in the livers of the DBA mouse and the cross between the C57BL/6J and the DBA mice, the B6D2F1/J mouse, there are little or no differences between the mouse strains examined and the Sprague-Dawley rat.

This comparison has been further expanded in a more recent study (15) which shows that the level of the receptor in the liver of the guinea pig is not significantly different from that in the Sprague-Dawley rat, the Macaca Fascicularis monkey, the C57BL/6J mouse, or the Golden Syrian hamster. In addition, the affinity of

TCDD for the receptor in these species did not appear to be significantly different. TCDD does not increase the activity of the _Ah_ locus enzymes, AHH, and DT-diaphorase in the guinea pig (16), which is the most sensitive species to the acute toxicity of TCDD. Thus, in contrast to the data in various mouse strains, there does not appear to be a correlation between enzyme induction and the presence of the TCDD-receptor in the guinea pig. Also, although the concentrations of receptors in the liver and the affinity of TCDD for these receptors are very similar in the various species, the acute toxicities are quite different. Recall, for example, that the LD_{50} of TCDD in the guinea pig is 2 ug/kg whereas in the hamster it is >3000 ug/kg (Table 1). A possible reason for the reduced acute lethality of TCDD in the hamster or for its inability to induce the enzymes of the _Ah_ locus in guinea pigs may be that the TCDD-receptor complex may not be transferred from the cytosol to the nucleus in these two species. However, data from the laboratory of Gasiewicz (personal communication) indicate that under the same _in vivo_ conditions, TCDD is translocated into the nucleus of the rat, mouse, hamster and the guinea pig in similar amounts. Thus, there are inconsistencies across species in the concept that the affinity of TCDD, and perhaps other chlorinated dibenzo-_p_-dioxins and dibenzofurans, for the cytosol receptor as well as the concentration of the receptor is related to the ability of these compounds to induce various enzymes and to cause other toxic effects including liver damage and acute lethality.

In spite of these inconsistencies, the data using various mouse strains (11-13,17) and the results of structure-activity studies (10) suggest that the binding of TCDD and related compounds to the receptor is in some way related to some of the biological effects of TCDD in experimental animals. There is no reason to believe that the alteration in the activity of enzymes noted in Table 2 is responsible for the toxic effects which are seen. A number of compounds can induce the activity of these same enzymes but not show the toxic effects of exposure to, for example, TCDD. One possible explanation, among others, for these data is that in addition to the binding to the receptor, TCDD causes additional biological effects which interfere with the normal functioning of the cell. The sensitivities of these effects vary with individual species. Studies with mammalian cells in culture (17-20) have shown that TCDD apparently has little or no effect on mammalian cell division or viability. These data also suggest that the toxicity of TCDD in whole animals may be related to the alteration of two or more biological parameters, one of which may be external to the affected cells.

Thymic involution is a consistent effect of TCDD in all animals so far examined. Shown in Table 4 are estimations of the concentrations of TCDD required to reduce the thymus weight by 50% in various species (15).

Table 4. Ability of TCDD to produce thymic atrophy in different species.

Species	ED_{50} (ug/kg)
Guinea pig	0.5 - 1.0
Rat	15
Mouse (C57BL/6J)	60
Hamster	>300

Gasiewicz (personal communication).

Note that the doses required to produce thymic atrophy in the guinea pig are much smaller than in the other species examined. However, the levels of the receptor in the thymus do not correlate with these ED_{50} values (15). Thus, the levels of receptor are higher in the rat thymus than in the guinea pig. Yet a lower dose of TCDD is required to cause thymic involution in the guinea pig. A further comparison of receptor concentrations in various tissues of the guinea pig, the most sensitive species to acute toxicity, with the hamster, the most resistant, shows that with the exception of the heart and testes, the concentrations are quite similar (15).

METABOLISM OF TCDD

Another question of interest is whether TCDD is metabolized and, if so, what is the effect of metabolism on the acute toxicity of TCDD? In other words, which compound is responsible for acute toxicity--the parent compound or a metabolite or metabolites?

Until recently, there was some question whether TCDD was metabolized in animals, particularly since it appears to be a poor substrate for soil bacteria. However, the work of Rose et al. (21), Poiger and Schlatter (22), and Olson et al. (23) have provided convincing evidence that TCDD is slowly metabolized in a number of species.

When hamsters are administered 500 ug/kg ^{3}H-TCDD and the urine collected for 24 hours starting on the 7th day following administration of TCDD, no parent compound was found to be excreted in the urine during this period (23). However, a number of compounds more polar than TCDD were found in the urine, some of which are apparently present as glucuronides (24). Also, some of the metabolites observed probably were ethereal sulfate derivatives of TCDD. An examination of the bile collected from these same animals again reveals no parent compound but instead a number of metabolites of TCDD, some of which appear to be glucuronide derivatives.

Incubation of ^{3}H-TCDD with primary hepatocytes isolated from hamsters and rats leads to the accumulation of a number of metabolites of TCDD in the incubation media, some of which appear to be glucuronides (24,25). Additional data indicate the presence of ethereal sulfate derivatives of TCDD. Similar results to these have also been obtained using hepatocytes isolated from rats, hamsters, and mice (24).

The major metabolites formed on incubation of primary rat hepatocytes with TCDD are 1-hydroxy-2,3,7,8-tetrachlorodibenzo-p-dioxin and 8-hydroxy-2,3,7-trichlorodibenzo-p-dioxin (25). These two metabolites represented about 60% of the metabolic products present in the incubation.

It is logical that the enzyme system responsible for the formation of the metabolites of TCDD seen on incubation with primary rat hepatocytes is the cytochrome P-450 monooxygenase system. In order to verify this, the effect of pretreatment of rats with the cytochrome P-450 inducer, phenobarbital (Pb), on the ability of the hepatocytes to metabolize TCDD was examined (24). It was found that pretreatment of rats with Pb markedly increased the rate of metabolism of TCDD by primary hepatocytes as compared to controls. Also, pretreatment with a small dose (5 ug/kg) of TCDD also markedly increased the rate of metabolism of TCDD by primary rat hepatocytes. When rat hepatocytes were incubated with TCDD in the presence of SKF 525-A (0.1mM) or metyrapone (0.5 mM), the metabolism of TCDD was inhibited. These data strongly suggest that the P-450 monooxygenase system is responsible for the metabolism of TCDD to the phenolic derivatives.

In order to assess whether the parent compound or metabolites were responsible for the acute toxicity, the LD_{50} of TCDD was determined in weanling rats and in weanling rats pretreated with Pb (50 mg/kg/3 days), 3-methylcholanthrene (3-MC) (40 mg/kg), or TCDD (5 ug/kg) (Table 5) (5).

Table 5. Toxicity of 2,3,7,8-tetrachlorodibenzo-p-dioxin (TCDD) to male weanling rats pretreated with phenobarbital (Pb), 3-methylcholanthrene (3-MC), or TCDD.

Treatment	LD_{50} (ug/kg, mean ± SE)
None	25.2 ± 1.4
Pb	40.9 ± 1.3[a]
3-MC	44.1 ± 1.2[a]
TCDD	36.8 ± 1.8[a]

[a] Significantly (P<0.05) different from controls (none).
Beatty, Vaughn and Neal (5).

In the rats pretreated with Pb, 3-MC or TCDD, the LD_{50} was increased relative to controls. These data suggest that metabolism leads to a decrease in the acute toxicity of TCDD and that the parent compound is probably responsible for the acute toxicity. This is reinforced by data from the work of Poiger and Buser (26) who administered the metabolites of ^{3}H-TCDD excreted in the bile of a dog to guinea pigs. On a molar basis (based on radioactivity), the metabolites in the bile were >100X less toxic than TCDD itself.

Shown in Table 6 are the half-lives for elimination of TCDD in various species compared with the LD_{50} of TCDD in those same species (27).

Table 6. Rates of elimination, AHH induction and toxicity of TCDD in various species.

Species	Dose (ug/kg)	Half-life for elimination (days)(t 1/2)	AHH induction	LD_{50} (ug/kg)	Reference
Guinea pig	2.0 (ip)	30	No	2	(27)
Rat	1.0 (oral)	31	Yes	60	(21)
Mouse:					
C57BL/6J	10.0 (ip)	17	Yes	132	(6)
DBA/2J	10.0 (ip)	37	Yes	620	(6)
B6D2F1/J	10.0 (ip)	17	Yes	300	(6)
Hamster	650 (ip)	11	Yes	>3000	(3)
Hamster	650 (oral)	15	Yes	5051	(3,4)

Gasiewicz, Olson, Geiger and Neal (27).

Note that the half-life for elimination of TCDD is the same in the guinea pig and rat, yet the LD_{50} values are quite different. Also, in the various mouse strains, the strain with the longest half-life for elimination, the DBA, is the least sensitive to the acute lethal effects. And, in the hamster, which is quite resistant to the acute lethal effects of TCDD, the half-life for elimination of TCDD is not greatly different from, for example, the C57/B6 mouse.

These data indicate the acute lethality of TCDD is apparently not directly related to the residence time of TCDD in the organism. And, since residence time is apparently related to the rate of metabolism of TCDD, the acute toxic effects probably are not directly related to the rate of metabolism of the compound.

ABSTRACT

2,3,7,8-Tetrachlorodibenzo-p-dioxin (TCDD) causes a variety of biological effects in experimental animals. The mechanism of the induction of certain enzymes is perhaps best understood: there is binding of TCDD to a receptor, translocation of the TCDD-receptor complex into the nucleus, followed by increased activity of a number of enzymes in the cell.

Although the concentration of this receptor in various tissues of some mouse strains correlates well with the intensity of some of the biological effects observed in the mouse strains exposed to TCDD, this correlation apparently does not extend across other species. The current evidence suggests that the acute toxic effects of TCDD in various species is in some way associated with binding of TCDD to the receptor. However, biological effects of TCDD in addition to those resulting from binding to the receptor may be required to produce acute toxicity and perhaps other effects.

The acute toxic effects of TCDD are probably caused by the parent compound rather than metabolites; however, this conclusion must be viewed as tentative. Also, it cannot be excluded at this time that biological effects other than acute toxicity may be caused by metabolites of TCDD. Finally, the acute toxic effects of TCDD appear not to be related, at least not directly, to the rate of metabolism of TCDD in experimental animals nor to the half-life of excretion.

REFERENCES

1. Poland, A., and Knutson, J.C. 2,3,7,8-Tetrachlorodibenzo-*p*-dioxin and related halogenated aromatic hydrocarbons: Examination of the mechanism of toxicity. Annual Review of Pharmacology and Toxicology 22:517-554 (1982).

2. McConnell, E.E., Moore, J.A., Haseman, J.K., and Harris, M.W. The comparative toxicity of chlorinated dibenzo-*p*-dioxins in mice and guinea pigs. Toxicology and Applied Pharmacology 44:335-356 (1978).

3. Olson, J.R., Holscher, M.A., and Neal, R.A. Toxicity of 2,3,7,8-tetrachlorodibenzo-*p*-dioxins in the Golden Syrian hamster. Toxicology and Applied Pharmacology 55:67-78 (1980).

4. Henck, J.M., New, M.A., Kociba, R.J., and Rao, K.S. 2,3,7,8-Tetrachlorodibenzo-*p*-dioxin: Acute oral toxicity in hamsters. Toxicology and Applied Pharmacology 59:405-407 (1981).

5. Beatty, P.W., Vaughn, W.K., and Neal, R.A. Effect of alteration of rat hepatic mixed-function oxidase (MFO) activity on the toxicity of 2,3,7,8-tetrachlorodibenzo-*p*-dioxin (TCDD). Toxicology and Applied Pharmacology 45:513-519 (1978).

6. Gasiewicz, T.A., Geiger, L.E., Rucci, G., and Neal, R.A. Distribution, excretion, and metabolism of 2,3,7,8-tetrachlorodibenzo-*p*-dioxin in C57BL/6J, DBA/2J, and B6D2F1/J mice. Drug Metabolism and Disposition 11:397-403 (1983).

7. Schwetz, B.A., Norris, J.M., Sparschu, G.L., Rowe, V.K., Gehring, P.J., Emerson, J.L., and Gerbig, C.G. Toxicology of chlorinated dibenzo-*p*-dioxins. Environmental Health Perspectives 5:87-99 (1973).

8. Nebert, D.W., and Gielen, J.E. Genetic regulation of aryl hydrocarbon hydroxylase induction in the mouse. Federation Proceedings 31:1315-1327 (1972).

9. Poland, A., Glover, E., and Kende, A.S. Stereospecific, high affinity binding of 2,3,7,8-tetrachlorodibenzo-p-dioxin by hepatic cytosol. Journal of Biological Chemistry 251:4936-4946 (1976).

10. Poland, A., Greenlee, W.E., and Kende, A.S. Studies on the mechanism of action of ten chlorinated dibenzo-p-dioxins and related compounds. Annals of the New York Academy of Sciences 320:214-230 (1979).

11. Neal, R.A., Olson, J.R., Gasiewicz, T.A., and Geiger, L.E. The toxicokinetics of 2,3,7,8-tetrachlorodibenzo-p-dioxin in mammalian systems. Drug Metabolism Reviews 13:355-385 (1982).

12. Poland, A., and Glover, E. 2,3,7,8-Tetrachlorodibenzo-p-dioxin: Studies on the mechanism of action. In: The Scientific Bases of Toxicity Assessment (H. Witschi, Ed.), Elsevier/North Holland Biomedical Press, New York, 1980, pp. 223-239.

13. Jones, K.G., and Sweeney, G.P. Dependence of the porphyrogenic effect of 2,3,7,8-tetrachlorodibenzo-p-dioxin upon inheritance of aryl hydrocarbon hydroxylase responsiveness. Toxicology and Applied Pharmacology 53:42-49 (1980).

14. Gasiewicz, T.A., and Neal, R.A. The examination and quantitation of tissue cytosolic receptors for 2,3,7,8-tetrachlorodibenzo-p-dioxin using hydroxylapatite. Analytical Biochemistry 124:1-11 (1982).

15. Gasiewicz, T.A. Receptors for 2,3,7,8-tetrachlorodibenzo-p-dioxin: Their inter- and intra-species distribution and relationship to the toxicity of this compound. In: Proceedings of the Thirteenth Conference on Environmental Toxicology, 16, 17, and 18 November 1982. Air Force Aerospace Medical Research Laboratory, Wright-Patterson Air Force Base, OH, (AFAMRL-TR-82-101) August 1983, pp. 250-269.

16. Neal, R.A., Beatty, P.W., and Gasiewicz, T.A. Studies of the mechanisms of toxicity of 2,3,7,8-tetrachlorodibenzo-p-dioxin (TCDD). Annals of the New York Academy of Sciences 320:204-213 (1979).

17. Beatty, P.W., Lemack, K.J., Holscher, M.A., and Neal, R.A. Effects of 2,3,7,8-tetrachlorodibenzo-p-dioxin (TCDD) on mammalian cells in tissue culture. Toxicology and Applied Pharmacology 31:309-312 (1975).

18. Kouri, R.E., Ratrie, H., Atlas, S.A., Niwa, A., and Nebert, D.W. Aryl hydrocarbon hydroxylase induction in human lymphocyte cultures by 2,3,7,8-tetrachlorodibenzo-p-dioxin. Life Sciences 15:1585-1595 (1974).

19. Knutson, J.C., and Poland, A. 2,3,7,8-Tetrachlorodibenzo-p-dioxin: Failure to demonstrate toxicity in twenty-three cultured cell types. Toxicology and Applied Pharmacology 54:377-383 (1980).

20. Niwa, A., Kumaki, K., and Nebert, D.W. Induction of aryl hydrocarbon hydroxylase activity in various cell cultures by 2,3,7,8-tetrachlorodibenzo-p-dioxin. Molecular Pharmacology 11:399-408 (1975).

21. Rose, J.Q., Ramsey, J.C., Mentzler, T.A., Hummel, R.A., and Gehring, P.J. The fate of 2,3,7,8-tetrachlorodibenzo-p-dioxin following single and repeated oral doses to the rat. Toxicology and Applied Pharmacology 36:209-226 (1976).

22. Poiger, H., and Schlatter, C. Biological degradation of TCDD in rats. Nature 281:706-707 (1979).

23. Olson, J.R., Gasiewicz, T.A., and Neal, R.A. Tissue distribution, excretion, and metabolism of 2,3,7,8-tetra chlorodibenzo-p-dioxin (TCDD) in the Golden Syrian hamster. Toxicology and Applied Pharmacology 56:78-85 (1980).

24. Olson, J.R., Gasiewicz, T.A., Geiger, L.E., and Neal, R.A. The metabolism of 2,3,7,8-tetrachlorodibenzo-p-dioxin in mammalian systems. In: Accidental Exposure to Dioxins (F. Coulston and F. Pocchiari, Eds.), Academic Press, New York, 1983, pp. 81-103.

25. Sawahata, T., Olson, J.R., and Neal, R.A. Identification of metabolites of 2,3,7,8-tetrachlorodibenzo-p-dioxin (TCDD) formed on incubation with isolated rat hepatocytes. Biochemical and Biophysical Research Communications 105:341-346 (1982).

26. Poiger, H., and Buser, H.R. Structure elucidation of mammalian TCDD-metabolites. In: Human and Environmental Risks of Chlorinated Dioxins and Related Compounds (R. Tucker, A. Young and A. Gray, Eds.), Plenum Press, New York, 1983, pp. 483-492.

27. Gasiewicz, T.A., Olson, J.R., Geiger, L.H., and Neal, R.A. Absorption, distribution and metabolism of 2,3,7,8-tetrachlorodibenzo-p-dioxin (TCDD) in experimental animals. In: Human and Environmental Risks of Chlorinated Dioxins and Related Compounds (R. Tucker, A Young and A. Gray, Eds.), Plenum Press, New York, 1983, pp. 495-525.

ANALYSES OF HUMAN TISSUES FOR CHLORINATED DIBENZO-*p*-DIOXINS AND CHLORINATED DIBENZOFURANS: THE STATE OF THE ART

Thomas O. Tiernan, Ph.D.*

Michael L. Taylor, Ph.D.*

Garrett F. VanNess*

John H. Garrett*

Steven R. Bultman*

Curtis B. Everson*

John D. Hinders*

Arnold Schecter**

The Brehm Laboratory and Department of Chemistry
Wright State University
Dayton, Ohio 45435

and

**Department of Preventive Medicine
Clinical Campus of the Upstate Medical Center
State University of New York at Binghamton
Binghamton, New York 13901

From Public Health Risks of the Dioxins, proceedings of a symposium held on October 19-20, 1983 at The Rockefeller University, New York City. Edited by William W. Lowrance. Published by William Kaufmann, Los Altos, California.

As a result of advances in analytical chemistry, particularly in the science of gas chromatography-mass spectrometry over the past five years or so, it is now possible to detect and quantitatively measure certain chlorinated dibenzo-p-dioxins (CDDs) and chlorinated dibenzofurans (CDFs) in human tissues at concentrations as low as parts per trillion. The fact that such compounds have indeed been found in tissues of humans apparently exposed to sources of these (Masuda et al., 1983; Rappe et al., 1983) is indicative of the human capacity to absorb these compounds from environmental sources (air, water, soil), although the principal route of human exposure in such instances has not yet been determined. Since the ultimate fate of these compounds within the human body following absorption and the possibility of their metabolism also have not been established, it is still not possible to quantitatively correlate the levels of these compounds detected in human tissues with exposure. While some qualitative correlations of CDDs/CDFs in various sources with corresponding compounds in human tissues have been made, accurate dose-response relationships for the CDDs/CDFs in humans have not been established. Indeed, even the nature of human symptoms and effects resulting from exposure to CDDs/CDFs are still topics of considerable controversy. It seems clear that progress in establishing human symptomatology and dose-response relationships for CDDs/CDFs in humans will require extensive collaboration between clinical medical investigators and analytical chemists to study appropriate human subjects. Hopefully, it will ultimately be possible, from the assessment of humans exposed to CDDs/CDFs in accidents such as chemical spills, fires, and explosions of chemical reactors, as well as in "normal" industrial working environments, to correlate observations of specific CDDs/CDFs and the quantities thereof in human tissues with resultant health effects. Obviously, if achieved, this will facilitate the understanding of the human response to CDDs/CDFs exposure, which is the necessary prelude to developing effective treatment procedures.

The following discussion focuses on the current capabilities and limitations of the analytical methods utilized to determine CDDs/CDFs in human tissues and presents some data relevant to recent human exposure episodes.

RATIONALE FOR CONSIDERING CDDs AND CDFs CONCURRENTLY

While it is possible in principle to analyze tissue samples for any of the discrete chlorinated dibenzodioxins or chlorinated dibenzofurans, and to treat these compounds in an entirely separate manner, there is considerable rationale for developing analytical procedures which will permit concurrent analyses of these two groups of compounds. As shown in Figure 1, the CDDs and CDFs are quite similar structurally, and the two groups of compounds are thought to elicit similar toxic effects.

Figure 1. Generalized chemical structures of chlorinated dibenzo-p-dioxins and dibenzofurans.

Moreover, CDDs and CDFs are known to coexist in various hazardous materials such as the products of pyrolysis of certain transformer oils (Schecter et al., in press, 1984), the effluents from municipal refuse incinerators (Taylor et al., 1983), and in other chlorinated hydrocarbon wastes which have been improperly dumped or buried at various sites throughout the U.S. Still another factor which argues for concurrent analyses of these two classes of compounds is that, because of their chemical similarity, the complex analytical procedures applicable for determination of both sets of these compounds are very similar and the analyses of both are readily combined.

ANALYTICAL METHODOLOGY FOR DETERMINING CDDs/CDFs IN HUMAN TISSUES

There are only a few laboratories in the U.S., Canada, Europe, and Japan which have successfully applied quantitative analytical procedures to measure the concentrations of CDDs and/or CDFs in human tissues. While the detailed procedures applied by these laboratories differ slightly in various respects, they utilize essentially the same analytical sequence, which entails sample digestion and extraction, preliminary fractionation utilizing chemical partitioning and liquid chromatography, and finally analysis of the sample extract, which has been enriched in the concentration of CDDs/CDFs, using coupled gas chromatography-mass spectrometry (GC-MS) for quantitation. In many respects, these methods are quite similar to those which have been implemented for other sample matrices (Tiernan, 1983).

The capabilities required of an acceptable analytical method for determining CDDs/CDFs in human tissue samples are dictated by several factors. First, certain of the CDDs/CDFs are extraordinarily toxic (Esposito et al., 1980). This indicates the need to measure these compounds at very low concentrations. In fact, in two recent reports of the finding of TCDD in human tissues, the observed concentrations were only a few parts per trillion (Young and Shepard, 1983; Ryan et al., in press, 1984), whereas in several more extensive exposures involving PCBs, parts-per-billion concentrations of chlorinated dibenzofurans were detected in various human tissues (Masuda et al., 1983). A second requirement of an acceptable analytical procedure is the ability to detect and quantitate discrete CDD and CDF isomers. As shown in Table 1, the total number of CDD isomers is 75, where there are 135 CDF isomers, and these include monochlorinated through octachlorinated compounds.

Table 1. Numbers of CDD and CDF isomers as a function of the number of chlorine substituents.

Number of Chlorine atoms	Number of CDD isomers	Number of CDF isomers
1	2	4
2	10	16
3	14	28
4	22	38
5	14	28
6	10	16
7	2	4
8	1	1
Total	75	135

It is known that the toxicities of these various isomers differ markedly (Esposito et al., 1980) and thus the capability to analyze for discrete isomers is important. The ability to accomplish this in a rigorous quantitative sense is presently somewhat limited, however, as discussed below. Still another capability required of an acceptable analytical method for human tissues is the ability to analyze a variety of tissue types. Since, as already noted, the disposition of CDDs/CDFs in the human body is not known, it is not clear which types of human tissues are most appropriate for analysis. In this situation, it is desirable to survey a variety of human tissues including blood, adipose, and various organ sections if these are available.

LIMITATIONS OF CURRENT ANALYTICAL PROCEDURES

The ability to analyze any samples for CDDs and CDFs depends upon the analyst having available some quantities of the pure CDD and CDF isomers which are to be determined. In order to develop a rigorous quantitative analytical procedure, the efficacy of any particular methodology must be demonstrated by showing that the method can detect the specific CDD/CDF isomers of interest. This requires that representative sample matrices (corresponding to the actual samples to be analyzed) be spiked with known quantitites of the CDD/CDF isomers to be measured, and that the recovery of these, when the analytical method is implemented, be determined. In addition, the response of the GC-MS instrument used to quantitate CDDs/CDFs must be calibrated by injecting known quantities of the pure CDD/CDF isomers. While many CDD/CDF isomers have been prepared in several different laboratories (Buser, 1975; Buser and Rappe, 1980; Lamparski and Nestrick, 1980; Lamparski and Nestrick, 1982; Buser and Bosshardt, 1978; Mazer et al., 1983; Taylor et al., 1984) the quantities of most of these isomers which have been produced in isolated form and in high purity have generally been inadequate for use in validating the analytical extraction and separations procedures or for calibrating GC/MS instrumentation. It must be realized, therefore, that most of the analytical data on the levels of various CDD and CDF isomers in tissues which have been reported thus far are at best "semi-quantitative," owing to these limitations in validating methodology.

The most extensive efforts to develop rigorous quantitative procedures for determining compounds such as the CDDs in various sample matrices have been directed toward 2,3,7,8-tetrachlorodibenzo-*p*-dioxin (TCDD). In this instance, a sufficient quantity of the pure compound has been prepared and isolated to permit development and validation of reliable quantitative analytical methods. In addition, isotopically-labelled 2,3,7,8-TCDDs have been prepared (both ^{13}C-labelled and ^{37}Cl-labelled) and these labelled TCDDs are routinely added to samples being analyzed for native TCDD content prior to extraction, in which cases they serve as internal reference standards for quantitative measurement of the native TCDD concentrations. Further, the recovery of the labelled standard serves as an indicator of the efficacy of the entire analytical procedure in such cases. The use of internal standards is the preferred method for quantitating native CDD and CDF isomers in any samples and has been shown in the case of 2,3,7,8-TCDD to yield much more reliable quantitative data than procedures which utilize external standards. However, as yet, relatively few isotopically-labelled CDDs and CDFs have been prepared in sufficient quantities to be widely available to analysts accomplishing these determinations. A few laboratories have acquired some isotopically-labelled standards, including ^{13}C- and/or ^{37}Cl-labelled 2,3,7,8-TCDD, 2,3,7,8-TCDF, OCDD, and 1,2,3,4,6,7,8-Hepta CDD. In addition, certain labelled Hexa-CDDs, Hexa-CDFs, and OCDF have recently been prepared by the Brehm Laboratory for use as internal standards in such analyses.

The ideal analytical procedure for determining CDDs/CDFs in human tissues would utilize at least one isotopically-labelled CDD and one labelled CDF for each chlorinated class (that is, for monochlorinated through octachlorinated CDDs/CDFs, a total of 16 internal standards) and would yield quantitative data for all 75 CDD and 135 CDF isomers. For the reasons discussed above, this is clearly not possible using even the best analytical methodology at the current state of development. It is possible, however, to obtain accurate quantitative information on the concentration of 2,3,7,8-TCDD in human tissues and this, of course, is of major interest because of the extraordinary toxicity of this isomer and the fact that it is apparently disseminated rather widely in the environment, as a result of improper disposal of chemical wastes containing this compound. It is also possible to obtain some information about the concentrations of certain other CDD/CDF isomers in tissue samples using state-of-the-art analytical methods, but the extent to which such information is definitive will depend heavily on the availability of appropriate CDD/CDF isomer standards for each analytical laboratory obtaining such data. Each analytical laboratory must demonstrate the ability to resolve and detect any given CDD or CDF isomer for which analytical data are reported in the presence of all the other CDD/CDF isomers of the same class. For instance, if the 1,2,3,6,7,8-hexa CDF is reported as being present in a given sample on the basis of observing the appropriate mass chromatographic response at the GC retention time corresponding to that of an authentic 1,2,3,6,7,8-hexa CDF standard, then it must also be demonstrated that this identification is unique by showing that this hexa-CDF isomer is gas chromatographically resolved from all the other 16 hexa-CDFs under the GC and MS conditions used for the analysis. If this is not done, the data can legitimately be reported only in terms of "1,2,3,6,7,8-hexa CDF and other possible co-eluting hexa-CDF isomers." As more CDD/CDF isomer standards become available, the capabilities to accomplish truly isomer-specific analyses will progressively improve, and the analytical methodology is presently in a continuing state of evolution toward that objective. A complete isomer-specific CDD/CDF analysis would be very costly, however, and would likely require several days of effort.

Among the practical limitations of currently available analytical methodology for determining CDDs/CDFs in tissues, cost is an important consideration. At present, the cost for these analyses typically ranges from $500 to $2000 per sample, depending upon the number of analytes measured. Implementation of such procedures requires the availability of highly skilled analytical chemists, a laboratory which is specially designed for handling toxic materials such as the CDDs/CDFs, and expensive and complex GC-MS instrumentation. Elaborate safety precautions are necessary in handling CDDs/CDFs, and analytical personnel must wear protective clothing including coveralls, gloves, masks and/or respirators, and shoe and head covers. Such analytical work must be accomplished in controlled-access laboratories which are equipped with appropriate air filtration and circulation equipment, glove boxes and hoods. The analytical methods described herein are labor intensive and analysis of a tissue sample for CDDs/CDFs typically requires 2-3 days, although a batch of several samples (typically as many as 8-10 samples) can be processed together.

Still another practical limitation of the analytical methods described herein is that the quantity of tissue required, provided that the attainment of parts-per-trillion detection limits is desired, is typically on the order of 2-5 grams. In some cases, it may be difficult to obtain such a quantity of tissue.

CURRENT BREHM LABORATORY PROCEDURES FOR DETERMINING CDDs/CDFs IN HUMAN TISSUES

The Brehm Laboratory of Wright State University has utilized several different analytical procedures for determining CDDs/CDFs in human tissue over the past few years. These methods differ somewhat in the sample extraction/digestion procedures and/or the preliminary fractionation procedures. As yet, there has been inadequate experience with these various methods to clearly indicate which is the method of choice. However, the alkaline digestion procedure appears to be superior for blood samples, whereas the acid digestion procedure appears to be better for adipose and organ tissues. The neutral extraction procedure is applicable to all types of human tissues, but may not completely dissolve the sample in all cases. The contact time of the acid or base with the sample and the strength of these are important because degradation of some higher chlorinated CDD/CDFs can occur if the digestion and subsequent acid-base washing sequences are not properly accomplished. In the method described below, alternative procedures are indicated at various points, and the analyst must utilize that particular procedure which proves to be most efficacious in terms of overall performance and recovery of internal standards for a given sample.

The gas chromatographic-mass spectrometric procedures described below are generally applicable for determining tetra- through octachlorinated CDD/CDFs. Of course, analyses can be limited to any particular chlorinated class of interest (for example, just tetra-CDDs) by monitoring only those ions which are indicative of that class. Mass spectrometric analysis of CDDs/CDFs is based on the presumption that these compounds exhibit unique mass spectral fragmentation patterns. Electron impact mass spectra of CDDs/CDFs exhibit intense molecular ions $(M)^+$ with the appropriate ion clustering which is caused by the two chlorine isotopes (^{35}Cl and ^{37}Cl). These compounds fragment principally to give $(M-COCl)^+$ and $(M-COCl-Cl_2)^+$ ions. Smaller intensities of various other ions also appear in the spectra of both CDDs and CDFs. In most instances, monitoring the molecular ions of the CDDs and CDFs will easily distinguish these compounds from other chlorinated aromatic hydrocarbons that may be present in the sample extracts. However, some PCBs, as well as DDE, can fragment to yield ion masses that may interfere with detection of certain CDDs, unless high resolution mass spectral measurements are made. Also, the identification of CDFs may be complicated by the presence of chlorinated diphenyl ethers, since the latter compounds yield intense (M^+-Cl_2) ions that have the same exact mass and number of chlorine atoms as the corresponding CDFs. In such cases, successful analysis depends upon the ability of the preliminary liquid chromatographic separation procedures and/or the gas chromatographic fractionation to separate such interfering compounds from the CDDs/CDFs extract, before the latter is introduced into the mass spectrometer.

While it is usually desirable in GC/MS analyses to obtain complete mass spectral scans to identify the analytes of interest, this is not feasible in analyses of human tissues, such as those described here, where one desires to achieve very low detection limits (for example, parts per trillion) for the CDDs/CDFs that correspond to detection and quantitation of a few picograms of the analyte. In this instance, typically only a few ions in the mass spectrum of each CDD/CDF to be analyzed are monitored, using a technique termed selected-ion monitoring (SIM). As each CDD or CDF isomer (or groups of these in instances where complete gas chromatographic resolution is not achieved) elutes from the gas chromatograph, the data system adjusts the mass spectrometer to monitor the appropriate set of ions for detection of that isomer. Obviously, a new set of ions is monitored for each chlorinated class of CDDs/CDFs, as progressively higher chlorinated isomers elute from the gas chromatograph.

The procedures outlined below are generally applicable for measuring parts per trillion (picograms of analyte per gram of sample) to parts per million (micrograms of analyte per gram of sample) of CDDs/CDFs in human tissues.

A. Digestion and Extraction of Sample

1. Place an accurately weighed aliquot (1-10 grams) of tissue (adipose, whole blood, blood serum, or organ section) into a 125 mL or 250 mL flint glass bottle (Teflon-lined screw cap) and add internal standards. (Typically 1-2 ng each $^{13}C_{12}$-2,3,7,8-TCDD and $^{13}C_{12}$-2,3,7,8-TCDF, and 4 ng each of $^{13}C_{12}$-OCDD and $^{13}C_{12}$-OCDF are added as a minimum. Other labelled internal standards may be added if these are available).

2. Utilize one of the three following alternative digestion/extraction procedures:

a. Add 75 mL of concentrated hydrochloric acid to the sample bottle and place the sample bottle in an ultrasonic bath for 2 hours. Remove the sample bottle from the ultrasonic bath and add 30 mL hexane to the sample. Place the sample bottle on a wrist-action shaker and agitate for 1 hour. Remove the sample bottle from the shaker (if emulsions are present, centrifuge the sample to eliminate them). Transfer the organic layer to another 125 mL flint glass bottle (Teflon-lined screw cap). Repeat extraction two additional times by adding 15 mL of hexane to the sample and agitating the bottle on the wrist-action shaker for 10 minutes, and then combining the organic layers.

b. Add 40 mL of 40% (w/v) potassium hydroxide and 20 mL of ethanol to the sample bottle with the tissue, place the sample bottle in an oven at 100°C on a wrist-action shaker, and agitate for 16 hours. Transfer the entire contents to a 250 mL flint glass bottle (fitted with a Teflon-lined cap) along with 60 mL of 1:1 (v/v) ethanol: water. Add 40 mL of hexane to the sample bottle and agitate for 1 hour. Remove the sample bottle from the shaker (if emulsions are present, add 1-2 mL of ethanol to the sample bottle). Transfer the organic layer to another 125 mL flint glass bottle (Teflon-lined screw cap). Repeat extractions two additional times by adding 15 mL of hexane to the sample and agitating on wrist-action shaker for 10 minutes and then combine the organic layers. Wash the organic sample extract with 40 mL doubly-distilled water by shaking it for 1 minute and removing and discarding the aqueous layer.

c. In this case, the sample should be placed in a 250 mL flint glass bottle prior to addition of the internal standards. Add 120 mL of 33% (v/v) hexane-in-acetone. Place the sealed sample bottle on a wrist-action shaker and agitate it for 1 hour. Add 40 mL of doubly-distilled water to the sample and agitate the sample bottle for 10 seconds. Transfer the organic layer to another 250 mL flint glass bottle (Teflon-lined screw cap). Repeat the extractions two additional times by adding 30 mL of hexane to the sample and agitating on a wrist-action shaker for 10 minutes and then combine the organic layers. Wash the organic extracts with 40 mL of doubly-distilled water by agitating with the extract for 1 minute, and then remove and discard the aqueous layer.

3. Wash the organic extract with 30 mL of concentrated sulfuric acid by agitating with the sample for 1 minute. Allow sufficient time for the aqueous and organic layers to separate completely, and then remove and discard the acid layer.

4. Repeat the concentrated sulfuric acid wash by agitating the acidified sample for 5 minutes. Remove and discard the acid layer.

5. Wash the organic layer with 50 mL doubly-distilled water, agitating for 1 minute, and again remove and discard the aqueous layer.

6. Wash the organic extract with 30 mL of aqueous potassium hydroxide (20% w/v) agitating for 10 minutes. Remove and discard the aqueous layer.

7. Repeat Step 5.

8. Repeat the concentrated sulfuric acid wash adding 30 mL of sulfuric acid to the sample extract and agitating the acidified sample for 10 minutes.

9. Repeat Step 5.

10. Add 5 g of anhydrous sodium sulfate to the organic extract and allow the mixture to stand for at least 15 minutes.

11. Quantitatively transfer the organic extract using hexane to a clean test tube and reduce the volume to approximately 5 mL by using a stream of prepurified nitrogen while maintaining the test tube at 55°C in a water bath.

B. Cleanup and Preliminary Fractionation of Crude Sample Extracts

1. Pack a Pasteur pipette (5 mm I.D. x 146 mm) with a plug of glass wool (prewashed with methylene chloride), followed by a 50 mm section of Florisil PR, 60/100 (≅1.0 g), and a second glass wool plug.

2. Activate the Florisil column at 140°C for 16 hours and store in an oven until just prior to use.

3. Remove the Florisil column from the oven and allow it to cool for 5 minutes. Pre-wash the column with 5 mL of methylene chloride, followed by 2 mL of 2% (v/v) methylene chloride-in-hexane. Discard these washings.

4. Transfer the entire sample extract, in 1 mL of hexane, to the Florisil column, using a pipette. Rinse the test tube which contained the extract with 2 mL of 2% (v/v) methylene chloride-in-hexane and also transfer the rinsate onto the Florisil column.

5. Repeat the rinses of test tube four additional times, using 2 mL aliquots of 2% (v/v) methylene chloride-in-hexane each time

and transfer these rinsings to the Florisil column. Discard the resultant column eluant.

6. Elute the chlorinated dioxins and furans with 8 mL of methylene chloride into a 15 mL test tube. Retain this fraction for further cleanup.

7. Concentrate the eluant in the test tube to approximately 1 mL, using a stream of prepurified nitrogen while the tube is immersed in a water bath maintained at 55°C.

8. Quantitatively transfer the organic eluant to a 3 mL micro-reactor using methylene chloride. Concentrate the extract to near dryness.

9. Construct a disposable liquid chromatography column as follows. Cut off a Pyrex 10 mL disposable pipette at the 4 mL mark and use the lower portion of the pipette. Pack the small end with a plug of silanized glass wool. Next add 2.8 grams of Woelm basic alumina, previously activated overnight at 600°C in a muffle furnace, and placed in a desiccator for 30 minutes just prior to use.

10. Using a disposable pipette, transfer the sample onto the liquid chromatography column.

11. Rinse the centrifuge tube with two consecutive 0.5 mL portions of hexane, and transfer the rinses to the alumina column.

12. Elute the column with 10 mL of hexane and discard the eluant (taking care not to let the column run dry).

13. Elute the column with 15 mL of 20% (v/v) methylene chloride-in-hexane and discard the eluant.

14. Elute the column with 15 mL of 50% (v/v) methylene chloride-in-hexane and retain the eluant.

15. Concentrate the solution to approximately 1 mL, using a stream of prepurified nitrogen as before. Rinse the centrifuge tube wall with an additional 1 mL of methylene chloride and reconcentrate.

16. Quantitatively transfer the residue (using methylene chloride) to a 3mL micro-reaction vessel.

17. Evaporate the solution in the micro-reaction vessel almost to dryness as previously, rinse the walls of the vessel with approximately 0.5 mL methylene chloride, and evaporate the contents just to dryness, storing the extract in the freezer until analysis.

18. Approximately 1 hour before GC-MS (GC-LRMS or GC-HRMS) analysis, dilute the residue in the micro-reaction vessel with an appropriate quantity of tridecane. Gently swirl tridecane on portion of vessel to ensure dissolution of the dioxins.

C. GC-MS Analysis of Cleaned-Up Sample Extracts

1. Parameters for the GC/MS/DS System:

a. Gas Chromatograph
i. Injector: Configured for capillary column, splitless/split injection (split flow on 60 seconds following injection; injector temperature, 300°).
ii. Carrier gas: Hydrogen, 30 lb head pressure.
iii. Column: 60M (0.25 mm I.D.) DB-5; temperature-programmed, see Table 2 for program. (Other columns can also be used including OV-101, Silar-10C, SP-2330, and Silov; each of these will resolve somewhat different CDD/CDF isomers.)

b. Interface
i. The capillary GC column can be directly coupled to the mass spectrometer ion source, or an intermediate enrichment device such as a jet separator can be used.
ii. Temperature of interface: 250°C.

c. Mass Spectrometer

i. Ionization: Electron impact (70 eV).

ii. Static resolution: 1:600 to 1:12,000 (10% valley), depending upon whether high resolution MS or low resolution MS is used.

iii. Source temperature: 250°C.

iv. Operating mode: Selected Ion Monitoring (SIM); for ions monitored see Table 2.

2. Calibration Procedures:
It is necessary to verify the performance and resolution characteristics of the capillary GC column utilized in the analyses by injecting an appropriate mixture of CDD/CDF isomer standards. This must be done on a regular basis (usually daily) because the performance of these GC columns degrades over time. The useful lifetime of such columns is not predictable and depends on both initial quality (coating efficiency, stationary phase) as well as the history of use (number of samples injected, maximum temperature to which subjected, relative cleanliness of sample extracts analyzed). Calibration of the GC-MS-DS system to accomplish quantitative analysis of CDDs/CDFs is accomplished by analyzing a series of calibration standards in order to obtain a calibration plot. Each standard solution is prepared to contain the same concentration of each of the stable-isotopically labelled internal standard(s) utilized, but a different concentration of the native CDD or CDF to be determined. Injection of these calibration mixtures into the GC-MS yield response factors, which are then used to quantitatively calculate the concentrations of CDDs/CDFs in the tissue sample extracts.

3. Criteria for Identifying CDDs/CDFs in Unknown Samples from Mass Chromatographic Data:
GC-MS responses (that is, mass chromatographic responses) resulting from analysis of tissue sample extracts are identified as arising from specific CDDs or CDFs, provided that they satisfy the following criteria:

a. Mass spectral responses must be observed at the masses corresponding to the appropriate ions (see Table 2) for particular CDDs and/or CDFs and intensities of these ions, for a given isomer, must maximize simultaneously. In addition, the chromatographic retention times observed for each CDD/CDF signal must be correct relative to that of the appropriate stable-isotopically labelled internal standard. In cases where fragment ions are monitored, in addition to the molecular ions, the ion intensities of these ion masses (such as $M\text{-}COCl^+$) should maximize within ± 1 second of the time at which the molecular ions maximize.

b. The ratio of the intensity of the $(M)^+$ signals to that of the $(M + 2)^+$ signals must be within ±15% of the theoretical ratio (for example, m/z320/m/z322 = 0.77 in the case of TCDD).

c. The intensities of the observed ion signals are considered to be detectable if each exceeds the baseline noise by a factor of at least 2.5:1. The observed ion intensities are considered to be quantitatively measurable if each ion intensity signal exceeds the baseline noise by a factor of at least 5:1.

d. Further confirmation of the finding of a particular CDD/CDF can be obtained by monitoring fragment ions arising from COCl loss (m/z 256.933 in the case of TCDD, for example). An appropriate response should be observed at these masses if the compounds detected are indeed CDDs or CDFs.

e. For reliable detection and quantitation of CDFs it is necessary to monitor signals arising from chlorinated diphenyl ethers, which, if present, could give rise to fragment ions having ion masses identical to those of the CDFs. Accordingly, in Table 2 appropriate chlorinated diphenyl ether masses are listed, which should be monitored simultaneously with the CDF ion masses. Only when a signal for the diphenyl ether mass is not detected can the signal obtained for an apparent CDF be considered to be attributable entirely to that CDF.

Table 2. Sequence of operations in GC-MS-DS quantitation of CDDs/CDFs in extracts of human tissue samples.

Elapsed Time (min)	Event	GC Column Temperature (°C)	Temperature Program Rate (°C/min)	Ions Monitored by Mass Spectrometer (m/z)	Compounds Monitored
0.00	Injection, splitless	190			
1.00	Turn on split valve	190			
1.00	Begin temp. program to 220°C	190	5		
6.00	Open column flow to mass spectrometer	215	5		
7.00	Column temp. hold	220			
14.00	Start Tetra Program; sweep= 350 ppm; time/mass = 0.08 sec.	220		258.930	TCDD
				303.902	TCDF
				305.899	TCDF
				319.897	TCDD
				321.894	TCDD
				327.885	$^{37}Cl_4$-TCDD
				331.937	$^{13}C_{12}$-TCDD
				337.840	HxDPE
22.00	Stop Tetra Program	220			
22.50	Start Penta Program; sweep = 350 ppm; time/mass = 0.12 sec.			290.894	PeCDD
				337.863	PeCDF
				339.860	PeCDF
23.00	Begin temp. program to 235°	220	5	353.858	PeCDD
26.00	Column temp. hold	235		355.855	PeCDD
32.00	Stop Penta Program				
32.50	Start Hexa Program; sweep = 350 ppm; time/mass = 0.20 sec.	235		326.852	HxCDD
				375.821	HxCDF
				375.818	$^{13}C_{12}$-HxCDF
				389.816	HxCDD
				391.813	HxCDD
				401.856	$^{13}C_{12}$-HxCDD
				433.759	ODPE

Table 2. Sequence of operations in GC-MS-DS quantitation of CDDs/CDFs in extracts of human tissue samples. Continued.

Elapsed Time (min)	Event	GC Column Temperature (°C)	Temperature Program Rate (°C/min)	Ions Monitored by Mass Spectrometer (m/z)	Compounds Monitored
33.00	Begin temp. program to 250°	235	5		
36.00	Column temp. hold	250			
42.50	Stop Hexa Program	250			
43.00	Start Hepta Program; sweep = 350 ppm; time/mass = 0.30 sec.	250		358.819	HpCDD
				407.782	HpCDF
				409.779	HpCDF
				423.777	HpCDD
				425.774	HpCDD
				477.720	DDPE
53.00	Stop Hepta Program	250			
53.50	Start Octa Program; sweep = 350 ppm; time/mass = 0.30 sec.	250		396.773	OCDD
				441.732	OCDF
				443.740	OCDF
54.00	Begin temp. program to 270°	250°	5	453.772	$^{13}C_{12}$-OCDF
				457.738	OCDD
				459.735	OCDD
58.00	Column temp. hold	270°		469.779	$^{13}C_{12}$-OCDD
				471.776	$^{13}C_{12}$-OCDD
65.00	Stop Octa Program				
65.00	Begin temp. program to 300°	270°	5		
71.00	Column temp. hold	300°			
75.00	Cool column to 190°				

OBSERVATIONS OF CDDs/CDFs IN HUMAN TISSUES

Over the past five years or so, the Brehm Laboratory has analyzed a variety of human tissues for CDDs/CDFs. These analyses have mostly been accomplished sporadically in connection with different medical cases (and therefore different subjects) and not as part of a concerted investigation. Many of these human tissue samples have been analyzed only for TCDD, and specifically for 2,3,7,8-TCDD. A representative sample of such data is summarized in Table 3.

Table 3. Residues of 2,3,7,8-tetrachlorodibenzo-p-dioxin (TCDD) in human tissue as measured by GC-low resolution MS.

Subject	Type and weight of tissue analyzed	Concentration (ppt)	Minimum detectable concentration (ppt)	Recovery of $^{13}C_{12}$-2,3,7,8-TCDD*
OC-1	Adipose, 1.5 g.	N.D.**	25	98%
PH-1	Adipose, 4.8 g.	N.D.	15	88%
RB-1	Adipose, 2.0 g.	N.D.	15	68%
RB-2	Adipose, 2.0 g.	8	-	82%
RB-3	Adipose, 3.0 g.	N.D.	8	64%
RB-4	Adipose, 2.0 g.	N.D.	9	50%
RB-5	Adipose, 3.0 g.	8	-	75%
RB-6	Adipose, 8.0 g.	N.D.	7	77%
NH-1	Blood, 30 g.	N.D.	3	83%
NH-2	Blood, 17 g.	N.D.	5	92%
NH-3	Blood, 17 g.	N.D.	7	76%
NH-4	Blood, 30 g.	N.D.	5	94%
SR-1	Adipose, 5.0 g.	N.D.	10	85%
SR-1	Blood, 20 g.	N.D.	1	90%
SR-2	Adipose, 5.0 g.	N.D.	10	87%
SR-2	Blood, 20 g.	N.D.	2	94%

* Internal standard added to sample prior to analysis.
** N.D. indicates Not Detected at concentration in excess of the minimum detectable concentration cited.

In general, these data mirror the findings from the total group of such samples analyzed by the Brehm Laboratory thus far (approximately seventy to date). The conclusion from these results is that TCDD residues do not generally appear to be present in human tissues collected from living subjects, although the detection limits achieved for the human tissue samples analyzed by our laboratory vary from a low of 2 ppt to a high of 25 ppt, the average being about 10 ppt or slightly lower for TCDD. The detection limit achieved for these samples depends primarily upon the sample type and the quantity available for analysis, and upon the nature and quantities of other organic residues in the sample. In general, human blood samples are more readily analyzed to very low detection levels than are adipose and other tissues. As already noted, however, little is known about the disposition of TCDD and similar compounds in the human body, although it might logically be expected that such compounds would not persist long in the blood after being introduced into the human body and that these would tend to accumulate in fatty tissues in which they should be more soluble.

In considering the data described above, it must also be recognized that little or no definitive information was obtained regarding the living and working environments of the subjects or their possible exposures to sources of TCDD. Nevertheless, the results summarized here are in marked contrast to the recently reported data of Ryan et al., 1984, who found detectable levels of 2,3,7,8-TCDD in the adipose tissue of 24 out of 25 subjects surveyed. In the latter case, the tissues had all been obtained from older, deceased persons (although again the nature of their environmental exposure was unknown), and the observed concentrations of 2,3,7,8-TCDD in the Canadian tissue samples ranged from about 4 to 22 ppt, about half of these exhibiting levels in excess of 10 ppt.

In a more recent investigation of several human subjects who were believed to have possible exposure to CDDs/CDFs as a result of a PCB-transformer fire (Schecter et al., 1984), both adipose and blood samples were collected from the subjects (as well as from several control subjects) and these were analyzed for tetra-, penta-, hexa-, hepta-, and octa-CDDs and CDFs. Soot produced from the fire, which extensively contaminated the entire building, and which may subsequently have exposed cleaning personnel, firemen, and other workers present in the building following the fire, was also analyzed. While a complete isomer specific analysis of the soot for all of the CDDs/CDFs was not possible, the data obtained indicated that very large concentrations of all of the chlorinated classes of CDDs and CDFs (at least tetra- through octachlorinated) were present in the soot, at levels ranging up to several hundred parts per million. The concentrations of total CDFs in the soot (4761 ppm) exceeded the concentrations of the total CDDs (19 ppm) by a factor of about 250. There appeared to be at least 40 discrete CDF isomers present in the soot (and probably many more which were unresolved under the gas chromatographic conditions utilized), including 2,3,7,8-TCDF.

Table 4. Concentrations of higher chlorinated dibenzo-p-dioxins (CDDs) and dibenzofurans (CDFs) in soot produced by a PCB-transformer fire, in adipose tissues of human subjects exposed to the environment in which the fire occurred, and in adipose tissues of unexposed (control) humans.

	Concentrations in parts per trillion			
	Hepta-CDFs	Hepta-CDDs	Octa-CDF	Octa-CDD
Soot from building where fire occurred	$405x10^6$	$7x10^6$	$66x10^6$	$5x10^6$
Exposed subjects:				
S-1	1500	300	<150	400
S-2	500	<100	<90	<500
S-3	1900	500	<100	600
S-4	800	100	<150	700
S-5	2800	200	<50	600
S-6	100	500	<50	1000
S-7	7700	200	<50	500
Unexposed Control Subjects:				
C-1	150	200	<100	550
C-2	70	200	<30	800
C-3	1500	130	<50	400

The concentrations of hepta- and octachlorinated CDDs/CDFs detected in the soot sample are shown in Table 4. Also shown are the measured concentrations of these same compounds in the adipose tissues of persons who were exposed as a consequence of being present in the building either during or following the fire. Finally, Table 4 shows similar data for adipose tissues collected from several control subjects who had no known contact with the fire environment. It is noteworthy, from the data presented here, that while the adipose tissues of the exposed subjects do indeed contain some of the same higher chlorinated dioxins as the soot, these same compounds also were detected in the adipose tissues of the unexposed control subjects. Some of the "exposed" subjects (for example, S-5 and S-7) appear to exhibit substantially higher levels of some of these higher CDDs/CDFs than those found in the controls, but the significance of this is not clear because it is virtually impossible to quantitatively determine on an independent basis the extent of exposure of the several persons who were present in the fire environment. It also seems significant that the tetra- through hexachlorinated CDDs/CDFs could not be detected in any of the adipose tissues of the subjects listed in Table 4 (with typical detection limits of 10-30 ppt), although there were large concentrations of many of the latter compounds in the soot produced by the fire, as already mentioned. Blood samples from the "exposed" and "control" subjects listed in Table 4 were also analyzed for tetra- through octachlorinated CDDs/CDFs, but none of these compounds could be detected in the blood.

On the basis of the data presented in Table 4, it would be difficult to establish a direct connection between exposure of the human subjects to the PCB-transformer fire environment and the observed CDDs/CDFs residues in their tissues. In attempting to assess these data, however, it should be noted that these tissue analyses for CDDs/CDFs were only recently completed (in early 1983), more than two years after the fire. The only chemical residue measurements done on human subjects shortly after the fire were PCB determinations. The latter data, which are presented elsewhere (Schecter et al., 1984), appear to correlate approximately with the levels of hepta-CDFs reported here in Table 4. There also appears to be some correlation of certain clinical symptoms of these subjects with the chemical residue data (Schecter et al., 1984) although much more extensive studies would be required to permit definitive conclusions based on these observations.

DIRECTIONS OF FUTURE RESEARCH

It is possible to identify several objectives of future research aimed at quantitatively measuring the levels of CDDs/CDFs residues in human tissues, and determining the significance of these findings with respect to human health effects. First, more definitive information about the identities of discrete CDD and CDF isomers in such tissues is needed, and it will be possible to obtain such data as additional CDD and CDF isomers become available in sufficient quantities and purity for use as analytical standards. Collaborative testing of analytical methodology for human tissues by interlaboratory studies will then be desirable. Second, the levels of CDD/CDF residues in a representative sample of the general population must be determined if the significance of such residues in humans exposed to hazardous environments containing these is to be established. Ultimately, of course, it would be desirable to establish dose-response relationships for these compounds in humans. It is also important, however, to identify and measure human metabolites of CDDs/CDFs in humans (if such exist) in order to properly interpret the observation or failure to observe residues of the actual CDDs/CDFs. Finally, the eventual interpretation of the finding of any toxic chemical residues in humans in terms of resultant health effects will require close collaboration on such studies between chemists and clinical medical scientists. Many additional investigations of this type will be required to obtain a statistically significant data base.

REFERENCES

Buser, H.R. (1975). Analysis of polychlorinated dibenzo-p-dioxins and dibenzofurans in chlorinated phenols by mass fragmentography. Journal of Chromatography 114: 95-108.

Buser, H.R., Bosshardt, H.P. and Rappe, C. (1978). Formation of polychlorinated dibenzofurans (PCDFs) from the pyrolysis of PCBs. Chemosphere 1: 109-119.

Buser, H.R. and Rappe, C. (1980). High-resolution gas chromatography of the 22 tetrachlorodibenzo-p-dioxin isomers. Analytical Chemistry 52: 2257-2262.

Esposito, M.P., Tiernan, T.O. and Dryden, F.E. (1980). Dioxins, EPA 600/2-80-197, U.S. Environmental Protection Agency, Cincinnati.

Lamparski, L.L. and Nestrick, T.J. (1980). Determination of tetra-, hexa-, hepta-, and octachlorodibenzo-p-dioxin isomers in particulate samples at parts-per-trillion levels. Analytical Chemistry 52: 2045-2054.

Lamparski, L.L. and Nestrick, T.J. (1982). Microchlorination procedure for synthesis of higher chlorinated dibenzo-p-dioxins from [^{13}C]-2,3,7,8-tetrachlorodibenzo-p-dioxin. Analytical Chemistry 54:402-406.

Masuda, Y., Kuroki, H. and Nagayama, J. (1983). Polychlorinated dibenzofurans in tissues of patients with Yusho and their enzyme-inducing activities on aryl hydrocarbon hydroxylase. In Chlorinated Dioxins and Dibenzofurans in the Total Environment, Choudhary, G., Keith, L.H. and Rappe, C., editors, Butterworth Publishers, Boston, 375-384.

Mazer, T., Hileman, F.D., Noble, Roy W. and Brooks, J.J. (1983). Synthesis of the 38 tetrachlorodibenzofuran isomers and identification by capillary column gas chromatography/mass spectrometry. Analytical Chemistry 55: 104-110.

Rappe, C., Nygren, M. and Gustafsson, G. (1983). Human exposure to polychlorinated dibenzo-p-dioxins and dibenzofurans. In Chlorinated Dioxins and Dibenzofurans in the Total Environment, Choudhary, G., Keith, L.H. and Rappe, C., editors, Butterworth Publishers, Boston, 355-365.

Ryan, J.J., Williams, D.T., Lau, B.P.-Y. and Sakuma, T. (1984). Analysis of human fat tissue from the Great Lakes area for 2,3,7,8-tetrachlorodibenzo-p-dioxin and chlorinated dibenzofuran residues. In Chlorinated Dioxins and Dibenzofurans in the Total Environment, Vol. II, Choudhary, G., Keith, L.H. and Rappe, C., editors, Butterworth Publishers, Boston, in press.

Schecter, A., Tiernan, T.O., Taylor, M.L., VanNess, G.F., Garrett, J.H., Wagel, D.J. and Gitlitz, G. (1984). Biological markers after exposure to polychlorinated dibenzodioxins, polychlorinated dibenzofurans, and polychlorinated biphenyls. Part I. Residues of PCDDs, PCDFs, and PCBs in adipose of humans exposed to a transformer fire. In Chlorinated Dioxins and Dibenzofurans in the Total Environment, Vol. II, Choudhary, G., Keith, L.H. and Rappe, C., editors, Butterworth Publishers, Boston, in press.

Taylor, M.L., Tiernan, T.O., Garrett, J.H., VanNess, G.F. and Solch, J.G. (1983). Assessments of incineration processes as sources of supertoxic chlorinated hydrocarbons: concentrations of polychlorinated dibenzo-p-dioxins/dibenzofurans and possible precursor compounds in incinerator effluents. In Chlorinated Dioxins and Dibenzofurans in the Total Environment, Choudhary, G., Keith, L.H. and Rappe, C., editors, Butterworth Publishers, Boston, 125-164.

Taylor, M.L., Tiernan, T.O., Ramalingam, B., Wagel, D.J., Garrett, J.H., Solch, J.G. and Ferguson, G.L. (1984). Synthesis, isolation and characterization of the tetrachlorinated dibenzo-p-dioxins and other related compounds. In Chlorinated Dioxins and Dibenzofurans in the Total Environment, Vol. II, Choudhary, G., Keith, L.H. and Rappe, C., editors, Butterworth Publishers, Boston, in press.

Young, A.L. and Shepard, B.M. (1983). A review of on-going epidemiologic research in the United States on the phenoxy herbicides and chlorinated dioxin contaminants. Chemosphere 12: 749-759.

CHEMICAL ANALYSES OF ADIPOSE TISSUES

Christoffer Rappe, Ph.D.

University of Umeå
Umeå, Sweden

From Public Health Risks of the Dioxins, proceedings of a symposium held on October 19-20, 1983 at The Rockefeller University, New York City. Edited by William W. Lowrance. Published by William Kaufmann, Los Altos, California.

When we started to analyze human tissues in Sweden five years ago, we worked with blood plasma analyses mainly in occupationally-exposed people. We also worked with blood samples from the Yusho patients in Japan and Taiwan. Recently we have been involved in analyzing blood samples from firemen and cleaning personnel exposed in PCB fires. I am not going to discuss these later studies, except to mention that 2,3,7,8-tetra- and 2,3,4,7,8-pentadibenzofurans could be detected in blood at 0.1 parts per trillion (ppt), and that all the firemen that we have analyzed so far, now more than 40, have been completely negative.

Now we are working mainly with adipose tissue, although we have some limited experience with liver and kidney tissues. Adipose tissue is taken from the abdomen by the method described by Dr. Young (this symposium). Since we have access to all the tetra-, penta-, hexa-, hepta-, and octa- isomers of both the dioxins and furans, we have tried to optimize our cleanup method for all dioxins and furans. While most analytical techniques in the literature optimize for only the 2,3,7,8-TCDD, we are trying to optimize for all dioxins and furans so that we can see the whole series. Our cleanup method is a slight modification of that described by David Stalling which is based on column chromatography including using a charcoal column. [Reference: Stalling, D.L. et al. Residues of Polychlorinated Dibenzo-p-dioxins and Dibenzofurans in Laurentian Great Lakes Fish. In: Human and Environmental Risks of Chlorinated Dioxins and Related Compounds, R. Tucker, A. Young and A. Gray (eds.), pp. 221-240. Plenum Press, New York, 1983.] We are also using isomer-specific determination of the isomers using high-resolution gas chromatography and mass spectrometry.

Table 1 presents preliminary results on adipose tissue analyses in our laboratories in Umeå, Sweden. They represent part of a collaborative investigation with Dr. Hardell on his soft-tissue sarcoma patients and his lymphoma patients, and also unexposed people undergoing normal gall bladder surgery. Also shown are unpublished analyses, reprinted by permission, conducted by Dr. Ryan in Canada.

Table 1. Levels of chlorinated dibenzodioxins and dibenzofurans in adipose tissue (pg/g=ppt).

		Cl_4		Cl_5		Cl_6		Cl_7		Cl_8	
	N	D	F	D	F	D	F	D	F	D	F
Sweden	6	1.5	3	12	40	11	22	73	50	240	3
Binghamton (Rappe et al. analyses)	1	NA	8	NA	9	NA	ND	ND	ND	ND	ND
Canada Hospital (Ryan)	9	10	3	13	18	76	15	133	35	616	ND
Accidents (Ryan)	5	6	ND	3	9	34	11	121	28	628	ND

D=Dioxins
F=Furans

ND = not detected.
NA = not analyzed.

Isomers analyzed are always the toxic 2,3,7,8-substituted Cl_4, Cl_5, Cl_6 isomers.

We have the highest sensitivity working with negative chemical ionization (NCI), so these are the levels for these specific 2,3,7,8-tetrafuran, the 2,3,4,7,8-pentafuran, all the hexas, heptas, and octas. In the dioxins series, we have high sensitivity for the hexa-, hepta-, and octa-chlorinated compounds. NCI does not provide such high sensitivity for the tetra- and penta-dioxins; for these we have to use the electron impact (EI) technique. The major constituents are the octadioxins, heptadioxins, and heptafurans; and as Professor Tiernan also has reported in this symposium, we found very low levels of the octafurans, much lower than the heptafurans and the octadioxins. We have not broken the sample code, so we do not know whether these people have been exposed, or whether they are cancer patients or controls.

When these data are compared with the Ryan studies from Canada (Table 1), we find very good agreement. The table gives the mean values for the first six samples analyzed; 12 additional samples have been analyzed since, and their profile is the same. We have few data on the tetra- and pentadioxins; we find the highest, as I indicated earlier, in the heptas and octas; the pentas are higher than the tetras; the hexas seem to have the same, or slightly lower values.

We also have one sample from Binghamton (Table 1): in that case, contrary to Dr. Tiernan's findings, we only find tetra- and pentafurans. Perhaps in the NCI we do not have such high sensitivity for the hexa-, hepta-, and octafurans as for the tetras. But the levels in this particular case are just what you would expect: 8 ppt for the 2,3,7,8-tetrafurans and 9 ppt for the 2,3,4,7,8-pentafurans.

We have observed that we always find the toxic 2,3,7,8-substituted tetra-, penta-, and hexa- isomers in our samples, in addition to the less toxic heptas and octas.

The conclusion we draw from these very preliminary data is that we have low ppt levels of tetra-, penta-, hexa-, and octa- dioxins and furans in the general background population. Normally, we find higher levels of furans than dioxins. In the case of the octa-substituted compounds, the octadioxins predominate over the octafurans. We have also observed the same isomers and the same relationships in these adipose samples as in environmental samples. We have analyzed fish samples from the Great Lakes and from the Baltic Sea and found the same profile and the same isomers in the human adipose tissue as we did in the environmental samples (mainly fish, but also birds from the Baltic).

If the dioxins and furans occur in the background population, how can we prove specific exposure--people exposed, for instance, in PCB fires, Times Beach, Seveso, Vietnam, or from consuming a heavy diet of fish? How can we differentiate between these types of exposure? I think the only way to do it is to use the pattern recognition ("cluster analysis") approach. We must analyze for all dioxins and furans, tetra- and higher, otherwise we will not know whether these low ppt levels represent general background or special exposure.

Now I will speculate on the source of this general background. As I indicated, in both the human adipose tissue and environmental samples, we had about the same levels of the 2,3,7,8-TCDD as the

1,2,3,7,8-pentadioxin, which is just about as toxic as the 2,3,7,8-TCDD. Many people conjecture that the source of the 2,3,7,8-TCDD could be 2,4,5-trichlorophenol (TCP), or hexachlorophene. But then, where would the 1,2,3,7,8-pentadioxin come from, which in some cases has higher levels than the 2,3,7,8-TCDD? We have analyzed many samples, and I have gone through the literature, but I have not found any study in which these particular isomers have been identified in any commercial product. While the chlorophenols do contain pentadioxins, we have no data indicating that it is the 1,2,3,7,8- isomer, so the question concerning chlorophenols remains. However, the 1,2,3,7,8-penta has been found in fly ash and other incineration products. Along with a whole series of furans and dioxins, these are the same isomers that have been found in environmental samples. Taking the data from Yusho patients, we know which isomers should be retained and which metabolized or excreted, and the isomers that we find in the fly ash can always be found in environmental samples.

I would urge that we not only look for 2,3,7,8-TCDD but also for the whole series of dioxins, furans, and perhaps even PCBs and other chlorinated compounds in adipose tissue.*

---

*Some of the technical points in this presentation are more fully explained in the critical review: Christoffer Rappe, Analysis of Polychlorinated Dioxins and Furans, <u>Environmental Science and Technology 18, No. 3</u>, 78A-90A, 1984.

ANALYSIS OF DIOXINS AND FURANS IN HUMAN ADIPOSE TISSUE

Alvin L. Young, Lt. Col., USAF, Ph.D.

Agent Orange Projects Office (10A7C)
U.S. Veterans Administration
Washington, DC 20420

From Public Health Risks of the Dioxins, proceedings of a symposium held on October 19-20, 1983 at The Rockefeller University, New York City. Edited by William W. Lowrance. Published by William Kaufmann, Los Altos, California, 1984.

A major concern in the conduct of epidemiologic studies of human populations assessing the impact of 2,3,7,8-tetrachlorodibenzo-p-dioxin (TCDD) is the lack of accurate exposure data. Indeed, as early as 1974 the United States Environmental Protection Agency (EPA) concluded that without such data, regulatory action could not be taken against dioxin-contaminated herbicides (11). As a consequence, EPA initiated the Dioxin Monitoring Program. Although EPA subsequently found TCDD in selected environmental samples, albeit infrequently, no TCDD was found in 103 milk samples from nursing mothers in three western states where dioxin-contaminated herbicides were routinely used (1). However, with continued development of sophisticated instrumentation for detecting both polychlorinated dibenzo-p-dioxins (PCDDs) and polychlorinated dibenzofurans (PCDFs) in picogram ($1x10^{-12}$) quantities, environmental monitoring has now resulted in confirmation that not only 2,3,7,8-TCDD but other dioxins and furans are widely distributed (6,12). Such findings suggest that humans may routinely be contaminated with these compounds.

Routine monitoring of human tissues for TCDD or indeed for other PCDDs or PCDFs should be undertaken only after a thorough evaluation of the analytical methods, and confirmation that TCDD accumulates in human tissues after known exposure. In 1973, Baughman and Meselson (3) developed an analytical procedure for the determination of TCDD in biological substrates. Although this method has undergone several modifications, it represents the basic approach to extraction and gas chromatographic-mass spectrometric (GC-MS) analysis for dioxins. Such a method modification was employed by Facchetti et al. (5) in 1978 in the determination of TCDD in the tissues of a person exposed to the toxic cloud at Seveso, Italy.

ANALYSES FROM SEVESO, ITALY

The study by Facchetti and co-workers was the first case of human exposure to TCDD in which an analysis was made of cadaveric tissue to detect and study the distribution of dioxin. The subject of the study was a 55-year old woman who had died from a pancreatic adenocarcinoma 7 months after the ICMESA accident in Seveso in July,

1976. Although the cancer was not a result of the exposure to TCDD, the woman was significantly exposed to the toxic cloud. During the passage of the toxic cloud the woman was eating a meal in her home with doors and windows open. In the four days after the event, the woman consumed vegetables from the garden attached to her home. Animals reared by the woman's family in an area adjacent to the home began to die over a period of 15 days after the event. The woman was evacuated from her home and the associated area after 16 days. Subsequent tests for TCDD indicated that the subject had lived in a sector of Zone A which had a mean soil concentration of 185 ug/m^2. On the basis of the circumstances recorded it was presumed that the woman absorbed toxic substances contained in the cloud by inhalation, ingestion and contact. Two young nephews of the subject living in the same building developed serious chloracne on the third and fourth days, respectively, after the exposure.

The results (means of three independent determinations) for the GC-MS analysis for 2,3,7,8-TCDD in selected human tissues are shown in Table 1.

Table 1. Tissue concentrations, parts per trillion, of 2,3,7,8-tetrachlorodibenzo-p-dioxin in the organs of a woman who died of cancer seven months after exposure, Seveso, Italy.[a]

Sample	2,3,7,8-TCDD/wet tissue
Fat	1840[b]
Pancreas	1040[bc]
Liver	150[bc]
Thyroid	85[c]
Brain	60[b]
Lung	60[b]
Kidney	40[b]
Blood	6[c]

[a] Data from Facchetti et al., 1981.

[b] Values obtained at a resolution of 2,500.

[c] Values obtained at a resolution of 10,000.

As noted, 2,3,7,8-TCDD was present in all of the tissues analyzed. On the basis of the concentrations observed it was possible to distinguish four groups of tissues: adipose tissue and pancreas tissue with levels between 1,000 and 2,000 ppt; liver tissue with levels between 100 and 200 ppt; other tissues (thyroid, brain, lungs, kidneys) with levels between 10 and 100 ppt; and blood with levels less than 10 ppt. The levels in the pancreas may have been abnormally high due to presence of the cancerous cells.

The data suggest that blood levels are one-tenth the level of TCDD in liver tissue and one one-hundredth the level in adipose tissue. Reggiani (9) estimated that the TCDD body burden in the above subject at the time of death was 40 ug. Because human milk has a high fat content where TCDD should reside, Reggiani (9) has also reported on TCDD levels in human milk from mothers exposed to TCDD while living in Zone A and B of Seveso at the time of and immediately after the release of the toxic cloud. These data are shown in Table 2. These data are in agreement with 1976 data from Baughman (2) on levels of TCDD and in human milk collected in 1970 from Vietnamese women living in areas sprayed with Agent Orange.

Table 2. TCDD levels, parts per trillion, in human milk from breast-feeding mothers of Zones A and B.[a]

Sample Number	TCDD Level (whole milk basis)
341	9.3
215	2.3
925	28.0

[a]Data from Reggiani, 1981.

ANALYSES ON MILITARY PERSONNEL

The issue of Agent Orange and the Vietnam veteran prompted the first major study of TCDD in human adipose tissue. Agent Orange, the major defoliant used in Vietnam, contained an average of about 2 ppm of 2,3,7,8-TCDD (13). It was suggested that TCDD might concentrate in the body's fat and long after exposure it would serve both as a cause of ill effects and as a marker to prove prior exposure to Agent Orange. Consequently the VA embarked on a small feasiblity study to test analytical methodology and to determine whether conclusions might be drawn regarding the significance of the results.

The study conducted by Hobson et al. (7) was initiated in 1979 with the selection of two groups of adult males: 1) twenty-one Vietnam veterans, all but two of whom claimed health problems related to Agent Orange exposure, and who volunteered for the fat biopsy; and 2) twelve veterans with no service in Vietnam. Ten of the latter group had no known exposure to any herbicides, were undergoing elective abdominal surgery, and volunteered to serve as controls. The other two individuals were active duty U.S. Air Force officers with known heavy and relatively recent exposure in connection with herbicide disposal operations.

The procedure called for the removal of 10-30 g of subcutaneous adipose tissue from the abdominal wall. This was accomplished surgically under local anesthesia. Precautions were taken before, during and after the procedure to avoid contamination by compounds such as hexachlorophene that could contain TCDD. Specimens were collected in glass containers previously rinsed with acetone and dried before use. All tissues were refrigerated during shipment to the assay laboratory. Each of the volunteers had a medical history, physical examination and routine clinical chemistry. The details of military service in Vietnam from the volunteer's report and his service record were examined to evaluate his potential exposure to herbicides using the dates, location and nature of his service. From these a rough estimate of the likelihood of exposure to TCDD was made without knowledge of the assay results.

The extraction and assay of all samples for TCDD were conducted at the University of Nebraska, Midwest Center for Mass Spectrometry, Lincoln, Nebraska. Gas chromatography/high resolution mass spectrometry was employed for quantification of 2,3,7,8-TCDD and coeluting isomers. Extracts that contained materials giving signals greater than 2.5 times noise at the exact mass of TCDD (that is m/z 321.8936 ± 0.0020) over the integration period were reanalyzed. For the second analysis, signal profiles of m/z 321.8936 and m/z 319.8965 were monitored over the elution period of 2,3,7,8-TCDD (determined by the injection of standard solutions). A positive detection was reported if signals were observed above the detection limit and if their intensity of ratio was consistent with the presence of four chlorine atoms in the molecule. Samples meeting all criteria except the isotope intensity ratio were considered to contain "not detectable" levels of TCDD. For these samples, it was judged that the presence of TCDD is not disproved by the observation of an incorrect isotope ratio at these low concentrations; rather, the presence of TCDD is not confirmed.

The results of this study are shown in Table 3. Fourteen of the 21 Vietnam veterans had levels of TCDD in their adipose tissue at or above the detection limit. Three of these men had detectable material that could not be validated as TCDD or the measured value was only questionably above the detection limit. Six Vietnam veterans had TCDD in amounts from 5 to 7 ppt. Three Vietnam veterans had TCDD in amounts from 9 to 13 ppt. One veteran had 63 and 99 ppt, and another had 23 and 35 ppt.

Table 3. 2,3,7,8-TCDD, parts per trillion, in human adipose tissue of Vietnam era veterans and military personnel.[a]

Total Number of Samples	Values at or above Detection Limit[b]	Range	Mean ± SD[c]
33	25	3-29, 99	7.7 ± 5.5 (24)
		Vietnam Experience	8.3 ± 6.9 (13)
		No Vietnam Experience	5.7 ± 3.1 (11)

[a]Data from Hobson et al., 1983.

[b]Detection limit of 2-6 ppt.

[c]Excluding the one outlying high sample.

Of the 12 individuals who had never served in Vietnam, five had TCDD identified in their fat (4,6,7,7, and 14 ppt). Six had values low enough to be considered equivocal or the detected material was not validated as TCDD. The remaining veteran had no detectable TCDD. In the two Air Force officers with known heaviest exposure, TCDD measured was never more than 3 ppt above the limit of detection.

Among the 21 Vietnam veterans there was no uniformity of symptoms, either immediately after exposure, at the time of biopsy, or during the intervening period. No one symptom or group of symptoms was common to veterans with detectible TCDD in their fat. The presence of TCDD did not mean ill health, nor did its absence indicate good health. No detailed statistical analysis of this small pilot series was attempted.

Hobson et al. (7) concluded that the results of the very complex and technically difficult analysis indicated that very low levels of TCDD, believed to be 2,3,7,8-TCDD, could be detected in human adipose tissue in the range 3-99 ppt. The levels, however, did not correlate well with known exposure and nonexposure, and there was no correlation with health status. The study results did indicate that the assay method was feasible, but would serve no clinically or administratively useful purpose until additional data are available on background levels of TCDD in the general United States population.

ANALYSES ON OHIO RESIDENTS

The Environmental Protection Agency has also analyzed human adipose tissue for TCDD. In 1981, Kutz (8) reported on six specimens of human adipose tissue collected from residents of an urban Ohio county to serve as control specimens for some analytical studies done by the EPA Dioxin Monitoring Program. These specimens were excised during post-mortem examinations from individuals with no recorded or known exposure to 2,4,5-T or silvex. Subsequently, they were analyzed in duplicate following the EPA Dioxin Monitoring Program protocol. Instrumental determinations were conducted at two independent laboratories.

The results, shown in Table 4, demonstrated that all specimens contained residues of 2,3,7,8-TCDD. Levels ranged between 5 and 12 ppt, with a detection limit below 5 ppt. Kutz emphasized that all studies conducted to date, including this one, have been accomplished utilizing small sample sizes and deliberate specimen selection criteria. Consequently, these few data cannot be construed as being representative of the general population.

Table 4. 2,3,7,8-TCDD, parts per trillion, in human adipose tissue collected in support of EPA's Dioxin Monitoring Program.[a]

Total Number of Samples	Number of Positives	Range
6	6	5-12

[a]Data from Kutz, 1981.

ANALYSES ON GREAT LAKES RESIDENTS

Recently, the Canadian scientists Ryan and Williams (10) released data on the analysis of human fat tissue from the Great Lakes area for 2,3,7,8-TCDD and dibenzofuran residues. The fat samples (10 to 20 g) were obtained from deceased elderly hospital patients from the communities of Kingston and Ottawa, Ontario. Ryan and Williams analyzed the fat tissues for the presence of 2,3,7,8-TCDD and furans by developing and modifying existing procedures which had been validated for other biological matrices. Particular attention was paid in the protocol to the preparation, solvent washing, and testing of glassware in order to minimize both high background noise from non-specific contaminants and false positives from TCDD residues when working in the picogram concentration. Blank reagent values averaged less than 1.0 pg per sample (about 0.5 ppt) with detection limits for fat samples varying between 1 and 4 ppt. The use of the internal standard, ^{13}C - 2,3,7,8-TCDD, was essential for quantitation with recoveries averaging 52.4 $\pm$ 18.6 percent (n=20).

The results obtained by Ryan and Williams are shown in Table 5. Levels of 2,3,7,8-TCDD were found in 22 of 23 samples analyzed. Values ranged from 4.1 to 130 ppt. Excluding the one outlying high sample, average values found were 10.7 ± 5.4 (n=22) with the highest value being 21.8 ppt. Grouping of the 22 samples which had been analyzed in a blind fashion with regard to origin showed that the 12 Kingston samples had an average of 12.4 ± 5.8 ppt (n=12) and the Ottawa samples 8.6 ± 4.4 ppt (n=10) but the difference was only significant at about the P = 0.1 level. No other peaks appeared in the chromatograms which satisfied criteria for other TCDD isomers. For comparative purposes, PCB values of these samples averaged 3.5 ± 4.2 (n=23) ppm with the Kingston samples higher than Ottawa (4.6 vs 2.1) but again this difference was not significant. A selected number of these fat samples also contained dibenzofuran residues. These data are shown in Table 6. Four of five were positive for 2,3,7,8-TCDF at an average of 2.8 ppt, and all six samples that were analyzed for 2,3,4,7,8-PCDF gave positive responses averaging 17 ppt and varying between 2 and 60 ppt. Ryan and Williams concluded that from these data it would appear that most human fat tissues from older patients in the Great Lakes area have low but measurable amounts of TCDD and PCDF.

Table 5. 2,3,7,8-TCDD, parts per trillion, in human adipose tissue from samples collected in the Great Lakes area.[a]

Total Number of Samples	Number Positive[b]	Range	Mean ± SD[b]
23	22	4.1-21.8, 130	10.7 ± 5.4(n=21)
		Kingston	12.4 ± 5.8 (12)
		Ottawa	8.6 ± 4.4 (9)

[a]Data from Ryan and Williams, 1983.

[b]Detection limit between 1 and 4 ppt.

[b]Excluding the one outlying high sample.

Table 6. Parts per trillion of two dibenzofurans in selected human adipose tissue from samples collected in the Great Lakes area.[a]

Furans	Total Number Samples	Number Positive	Range	Mean
2,3,7,8-TCDF	5	4	-	2.8
2,3,4,7,8-PCDF	6	6	2-60	17.0

[a]Data from Ryan and Williams, 1983.

CONCLUSIONS AND PROPOSED STUDIES

In assessing the results of the above reports on analyses for dioxins and furans, the following observations are noted:

1. Levels of 2,3,7,8-TCDD at the low part per trillion were present in 85 percent of all human adipose tissue samples analyzed.
2. Except for the samples analyzed by Facchetti et al. (15), minimal exposure data were available.
3. Interpretation of data from all of the studies reported was limited by small sample sizes.
4. With the exception of selected samples analyzed by Ryan and Williams (10), no data were available on PCBs, furans or other dioxins.

It is apparent that a large scale scientific study must be undertaken to determine the magnitude of the dioxins and furans in adipose tissue of the general population. This need has been recognized by the EPA in their August, 1983 Dioxin Strategy (4). Moreover, before the Veterans Administration can conclude whether adipose tissue analyses of Vietnam veterans would assist in documentation of exposure to Agent Orange, additional data are needed on background levels of 2,3,7,8-TCDD in the U.S. population (13).

Accordingly, the VA in cooperation with EPA, has agreed to study levels of 2,3,7,8-TCDD in adipose tissue from a selected group of males.

EPA has been collecting adipose tissue from the United States general population by region, age, sex and race. This National Adipose Tissue Bank was started in 1968 and now contains specimens from over 12,000 individuals. Adipose tissue from approximately 550 males born between 1937 and 1952 is available from this bank. Many of these men served in the military during the Vietnam era and some served in Vietnam during the period of Agent Orange use. A retrospective study of their adipose tissue may establish data on background levels of 2,3,7,8-TCDD in the male population, as well as whether service in the military and especially in Vietnam has had an effect on the levels of TCDD in adipose tissue.

The study will be conducted in three phases. In Phase I the name and social security number will be obtained for the approximately 550 males noted above. This information will be used to determine military service status. Phase II will be the development of analytic methods for the determination of selected dioxins (especially 2,3,7,8-TCDD) and furans in human adipose tissue. The proposed methods will be subjected to rigorous interlaboratory validation by an independent analytic referee, such as the Association of Official Analytical Chemists. Phase III will be the residue analyses of the adipose tissue and the preparation of a final report. Phases I and II should be completed by mid-1984 and the report from Phase III should be available in 1985.

REFERENCES

1. Anonymous. 1980. Dioxin not detected in mother's milk. EPA Environmental News, R-10, January 14, 1980, United States Environmental Protection Agency, Washington, DC, 2p.

2. Baughman, R.W. 1976. Tetrachlorodibenzo-p-dioxins in the environment. High resolution mass spectrometry at the picogram level. Dissertation Abstracts International 36(7):3380B.

3. Baughman, R.W. and M. Meselson. 1973. An analytical method for detecting TCDD (Dioxin): Levels of TCDD in samples from Vietnam. Environmental Health Perspectives 5:27-35.

4. Dioxin Strategy Task Force. August, 1983. Dioxin Strategy. United States Environmental Protection Agency, Washington, DC, 33p.

5. Facchetti, S., A. Fornari and M. Montagna. 1981. Distribution of 2,3,7,8-tetrachlorodibenzo-p-dioxin in the tissues of a person exposed to the toxic cloud at Seveso. Forensic and Environmental Application 1:1405-1414.

6. Harless, R.L., R.G. Lewis, A.E. Dupuy and D.D. McDaniel. 1983. Analysis for 2,3,7,8-tetrachlorodibenzo-p-dioxin residue in environmental samples. Environmental Science Research 26: 161-171.

7. Hobson, L.B., L.E. Lee, M.L. Gross and A.L. Young. 1983. Dioxin in body fat and health status: A feasibility study. Extended Abstracts, Division of Environmental Chemistry, American Chemical Society 23(2):91-93.

8. Kutz, F.W. 1981. Chemical Exposure Monitoring in the EPA Office of Pesticides and Toxic Substances. Prepared statement to the Veterans Administration Advisory Committee on Health Effects of Herbicides, November 19, 1981, Washington, DC, 10p.

9. Reggiani, G. 1981. Medical survey techniques in the Seveso TCDD exposure. Journal of Applied Toxicology 1(6):323-331.

10. Ryan, J.J. and D.T. Williams. 1983. Analysis of human fat tissue from the Great Lakes Area for 2,3,7,8-tetrachlorodibenzo-p-dioxin and -furan residues. Extended Abstracts, Division of Environmental Chemistry, American Chemical Society 23(2):157-158.

11. Shapley, D. 1974. Briefings: Score one for Dow. Science 185: 509.

12. Stalling, D.L., L.M. Smith, J.D. Petty, J.W. Hogan, J.L. Johnson, C. Rappe, and H.R. Buser. 1983. Residues of polychlorinated dibenzo-p-dioxins and dibenzofurans in Laurentian Great Lakes fish. Environmental Science Research 26:221-240.

13. Young, A.L., H.K. Kang, and B.M. Shepard. 1983. Chlorinated dioxins as herbicide contaminants. Environmental Science and Technology 17(11):530A-540A.

SUMMARY AND CRITIQUE OF RODENT CARCINOGENICITY STUDIES OF CHLORINATED DIBENZO-*p*-DIOXINS

Richard J. Kociba, D.V.M., Ph.D.

Toxicology Research Laboratory
Dow Chemical U.S.A.
Midland, Michigan 48640

From Public Health Risks of the Dioxins, proceedings of a symposium held on October 19-20, 1983 at The Rockefeller University, New York City. Edited by William W. Lowrance. Published by William Kaufmann, Los Altos, California.

This presentation is an overview of rodent studies pertaining to the carcinogenicity of the chlorinated dioxins with specific reference to the 2,3,7,8-tetrachloro isomer (TCDD). Hopefully this overview will serve as an appropriate lead-in to subsequent presentations that will cover the related aspects such as quantitative risk assessments that have been derived from these studies on carcinogenicity.

Table 1 lists the chlorinated dioxins for which we presently have or will be getting experimental data. There are carcinogenicity data from rodent studies on the unsubstituted isomer, on the 2,7-dichloro isomer, the 2,3,7,8-tetrachloro isomer and a mixture of two hexachloro isomers. The octachloro isomer has been nominated by the NTP for testing in their cancer bioassay.

Table 1. Animal studies on carcinogenicity of chlorinated dibenzo-p-dioxins.

Dioxin	Dose Level	Results/Reference
Unsubstituted Dibenzo-p-dioxin		
Rat	10,000 ppm in diet	No carcinogenic response
Rat	5,000 ppm in diet	No carcinogenic response
Mouse	10,000 ppm in diet	No carcinogenic response
Mouse	5,000 ppm in diet	No carcinogenic response (NTP, 1979a)
2,7-Dichlorodibenzo-p-dioxin		
Rat	10,000 ppm in diet	No carcinogenic response
Rat	5,000 ppm in diet	No carcinogenic response
Mouse	10,000 ppm in diet	Suggestive carcinogenic response
Mouse	5,000 ppm in diet	No carcinogenic response (NTP, 1979b)

Table 1. Continued.

Dioxin		Dose Level	Results/Reference
2,3,7,8-Tetrachloro-dibenzo-p-dioxin (See Tables 2 and 3 for data.)			
1,2,3,6,7,8 and 1,2,3,7,8,9 Hexachlorodibenzo-p-dioxins (mixture)			
	Rat	5, 2.5 or 1.25 ug/kg/week via gavage	No carcinogenic response in males; carcinogenic response in females at higher dose levels (NTP, 1980)
	Mouse	5, 2.5 or 1.25 ug/kg/week via gavage (males)	Carcinogenic response at high dose level (NTP, 1980)
		10, 5 or 2.5 ug/kg/week via gavage (females)	Carcinogenic response at high dose level (NTP, 1980)
Octachlorodibenzo-p-dioxin		(Nominated for Testing in NTP Bioassay Program.)	

The unsubstituted dibenzo isomer has been studied in both the rat and the mouse in conventional lifetime two-year studies. This material is relatively nontoxic compared with the others and thus the dosages used were relatively high - 10,000 and 5,000 parts per million - in the diet for the rat and mouse for two years. In both species there was no carcinogenic response. The 2,7-dichloro isomer was similarly studied at comparably high dose levels in both rodent species. In the rat there was no carcinogenic response; in the mouse there was a suggestive effect in the form of an increased incidence of hepatocellular neoplasms at the high dose level.

The studies on the two hexachloro isomers utilized oral gavage as the route of administration to rats and mice. For the rat, the NTP report stated that there was no carcinogenic response in male rats, but there was a carcinogenic response in female rats at the higher dose levels. In the mouse, there was a carcinogenic response at the high dose level. This particular study has been recently reviewed by Dr. Robert Squire (1983) who supplies additional perspective on the overall interpretation of this hexachloro mixture.

The 2,3,7,8-tetrachloro isomer is the one for which we have the most information. Table 2 depicts data in a decreasing dosage schedule from the two definitive studies in rats that have been reported. There is a good concurrence between results of these studies. The dosage range of 0.07 to 0.1 ug/kg/day elicited a positive carcinogenic response in the Sprague-Dawley and Osborne-Mendel strains of rats. An intermediate dosage range of 0.007 to 0.01 ug/kg/day elicited a less than definitive carcinogenic response in the rat. At the lower dosage range of 0.001 to 0.0014 ug/kg/day for a lifetime, both rat studies report no increase in tumors.

Table 2. Carcinogenic assessment of TCDD in rats.

Calculated Daily Dose TCDD (ug/kg/day)	Rat Strain	Response/Reference
0.1	S-D	Hepatocellular carcinoma, Squamous carcinoma of oropharynx and lung (Kociba et al., 1978)
0.07	O-M	Hepatocellular carcinoma, Thyroid tumors (NTP, 1982a)
0.01	S-D	Hepatocellular nodules (Kociba et al., 1978)
0.007	O-M	Questionable increase in thyroid tumors (NTP, 1982a)
0.014	O-M	No increase in tumors (NTP, 1982a)
0.001	S-D	No increase in tumors (Kociba et al., 1978)

S-D = Sprague-Dawley
O-M = Osborne-Mendel

Table 3 lists data from the two long-term studies on the 2,3,7,8 isomer in mice. Evaluation of the dose response indicates good concurrence: higher dosages were associated with an increase in hepatocellular tumors in both the B6C3F1 mouse and in the Swiss mouse. At dosages of 0.03 ug/kg/day and lower, both studies again concur in showing no increase in tumors.

Table 3. Carcinogenic assessment of TCDD in mice.

Calculated Daily Dose TCDD (ug/kg/day)	Mouse Strain	Response/Reference
1.0	Swiss (M)	No increase in tumors, but decreased lifespan (Toth et al., 1979)
0.3	B6C3F1 (F)	Hepatocellular tumors/ thyroid tumors (NTP, 1982a)
0.1	Swiss (M)	Hepatocellular tumors (Toth et al., 1979)
0.07	B6C3F1 (M)	Hepatocellular tumors (NTP, 1982a)
0.03	B6C3F1 (F)	No increase in tumors (NTP, 1982a)
0.007	B6C3F1 (M)	No increase in tumors (NTP, 1982a)
0.006	B6C3F1 (F)	No increase in tumors (NTP, 1982a)
0.0014	B6C3F1 (M)	No increase in tumors (NTP, 1982a)
0.001	Swiss (M)	No increase in tumors (Toth et al., 1979)

Table 4 summarizes key studies that have addressed the issue of tumor initiation, tumor promotion and cocarcinogenesis. Whereas the earlier studies gave a somewhat mixed response, more recent studies indicate that TCDD carcinogenesis appears to act via a promoter mechanism.

Table 4. Studies with TCDD of tumor initiation, promotion, and cocarcinogenesis.

Test System	Results/Reference
TCDD on mouse skin with TPA as promoter	TCDD a weak initiator (Di Giovanni et al., 1977)
TCDD + DMBA on mouse skin with TPA as promoter	TCDD a rather inactive cocarcinogen (Di Giovanni et al., 1977)
TCDD + BAP on mouse skin with TPA as promoter	TCDD inhibited BAP tumor initiation (Cohen et al., 1979)
TCDD + DMBA on mouse skin with TPA as promoter	TCDD inhibited DMBA tumor initiation (Cohen et al., 1979)
DMBA on mouse skin with TCDD as promoter	TCDD not a promoter of skin tumors (Berry et al., 1978)
TCDD given SQ or IP prior or with MCA given SQ to mice	Mixed response; no clear-cut results (Kouri et al., 1978)
TCDD orally after DEN + partial hepatectomy of rats	TCDD a promoter of liver tumors (Pitot et al., 1980)
DMBA + MNNG on hairless mouse skin with TCDD as promoter	TCDD a promoter of skin tumors in hairless mice (Poland et al., 1982)

TPA = 12-0-Tetradecanoylphorbol-13-acetate.
DMBA = Dimethylbenzanthracene.
BAP = Benzo-a-pyrene.
MCA = Methylcholanthrene.
DEN = Diethylnitrosamine.
MNNG = Methyl-N-Nitrosoguanidine.

Table 5 summarizes one of these recent studies (Poland et al., 1982) in which the HRS/J strain of hairless mouse was used. This mouse is one of the few species that shows the chloracne-like proliferation of the integumentary epithelial tissue, similar to the response of the rabbit ear as a model of the chloracnegenic response in human species. Using this hairless strain of mouse that is genotypically heterogeneous for the hairless trait (phenotypically haired), TCDD does not promote skin tumors initiated by Dimethylbenzanthracene (DMBA). Using mice that are genotypically homozygous (phenotypically hairless), TCDD does promote skin papillomas initiated with either DMBA or Methyl-N-Nitrosoguanidine (MNNG). Those researchers concluded that the skin effects of TCDD in this hairless strain of mouse depend on the interaction of two genetic loci: first, the Ah locus which is associated with the cytosol receptor in its enzyme induction; and second, the hr locus which dictates whether or not epidermal proliferation will occur. They also concluded that their data supported a promoter mechanism of TCDD carcinogenesis.

Table 5. TCDD promotion of skin tumors in hairless mice.

HRS/J Mice	Initiator	Promoter	Skin Papillomas
Hairless (hr/hr)	DMBA	TPA	Yes
	MNNG	TPA	Yes
	DMBA	TCDD	Yes
	MNNG	TCDD	Yes
Haired (hr/+)	DMBA	TPA	Yes
	DMBA	TCDD	No

TPA = 12-0-Tetradecanoylphorbol-13-acetate.
DMBA = Dimethylbenzanthracene.
MNNG = Methyl-N-Nitrosoguanidine.

(after Poland et al., 1982)

A recent and more pertinent study is the liver tumor promotion study by Pitot et al. (1980). This group utilized a test system wherein intact rats are subjected to partial hepatectomy in order to stimulate hepatocellular proliferation followed by subsequent systemic dosing with the testing material for initiation and/or promotion. In this particular study, partial hepatectomy followed by TCDD alone did not initiate the liver foci or liver tumors (Table 6). Similarly, phenobarbital alone did not initiate the liver foci or tumors. The initiating agent diethylnitrosamine (DEN) did initiate hepatocellular foci. The group of rats initiated with DEN followed by a lower dose of TCDD promoted liver foci but no liver tumors. The group of rats initiated with DEN and given a higher dosage of TCDD did lead to promotion of liver foci and liver tumors. The group of rats promoted with phenobarbital gave similar results, and these authors concluded that this demonstrated the promoter mechanism of TCDD carcinogenesis in this particular test system. They stated it mimicked phenobarbital as a liver tumor promoter, but TCDD was more potent than phenobarbital in this particular test system.

Table 6. TCDD promotion of liver tumors in rats.

	Liver Foci	Liver Tumors
PH + TCDD	No	No
PH + Phenobarbital	No	No
PH + DEN	Yes	No
PH + DEN + TCDD (low dose)	Yes	No
PH + DEN + TCDD (high dose)	Yes	Yes
PH + DEN + Phenobarbital	Yes	Yes

PH = Partial hepatectomy.
DEN = Diethylnitrosamine.

(after Pitot et al., 1980)

An even more recent study that warrants discussion is the two-year skin painting study with TCDD done by the National Cancer Institute as part of the National Toxicology Program (NTP, 1982b). In this study, female mice showed an increased incidence of fibrosarcoma at the topical site of application of TCDD for the two-year treatment period. Table 7 lists the incidence of fibrosarcoma at the local site of skin application; it also lists the incidence at which inflammation/necrosis and ulceration of the subcutaneous tissue occurred.

Table 7. Two-year skin-painting study with TCDD.

	Untreated Control	Vehicle Control	TCDD	DMBA+ TCDD
Female Mice				
Fibrosarcoma at local site of application	1/27	2/41	8/27	8/29
Inflammation/necrosis of subcutaneous tissue	1/27	2/41	5/27	6/29
Ulceration into subcutaneous tissue	0/27	1/41	6/27	0/29
Male Mice				
Fibrosarcoma at local site of application	0/28	3/42	6/28	5/30
Inflammation/necrosis of subcutaneous tissue	0/28	4/42	3/28	4/30
Ulceration into subcutaneous tissue	0/28	1/42	2/28	2/30

DMBA = Dimethylbenzanthracene.

(after NTP, 1982b)

The fibrosarcomas reportedly induced by treatment were limited to the local site of topical application, and were preceded or accompanied by inflammation/necrosis and ulceration of the underlying subcutaneous tissue. As mouse integument is rather delicate and at most two or three epithelial layers thick, it appears that repeated lifetime application of some materials to this thin and delicate epithelial covering can lead to denudation and/or ulceration of the superficial epithelium. Repeated application would likely result in the material being directly applied to the underlying mesenchymal tissue. This mesenchymal tissue has only a limited number of ways of responding to insults, and one of these responses is the proliferation of the mesenchymal type of tissue leading to fibrosarcoma.

In this NTP study (1982b), the authors stated there was no evidence that TCDD was a systemic tumorigen when applied to the backs of mice, and they urged caution in trying to interpret the findings of an increased incidence of localized subcutaneous fibrosarcoma in the female mice. The NTP report also stated that "the tumor type is induced by implanted inert plastics." Historical research has documented the relative ease in which one can induce this fibrosarcoma type of reaction within both the rat and mouse subcutaneous tissue with a number of rather innocuous materials (Grasso and Goldberg, 1966).

Additional data regarding the possible mechanism of carcinogenesis in animal studies are listed in Tables 8 and 9. Table 8 summarizes the results from six groups of researchers that have used the Ames-Salmonella bacterial test system for evaluating mutagenic base pair substitutions as well as frame shifts. For the majority of these Salmonella test strains, TCDD has been negative for mutagenic activity. The one exception is with strain TA1532 in which two of four groups have reported TCDD to be positive for a frame shift mutation.

Table 8. Mutagenic studies with TCDD in salmonella bacterial tests.

Tester Strain	Study					
	A	B	C	D	E	F
Base Pair Substitutions						
G-46				-		-
TA1530			-	-		-
TA1535	-	-			-	-
TA100					-	-
TA1950						-
TA1975						-

Table 8. Continued.

Tester Strain	Study A	B	C	D	E	F
Frame Shifts						
TA1531				?		
TA1532	-		+	+		-
TA1534				?		
TA1537	-				-	-
TA1538	-	-			-	-
TA98					-	-
TA1978						-

Study A = McCann (as cited by Wassom et al., 1977-78).
Study B = Nebert et al., 1976.
Study C = Hussain et al., 1972.
Study D = Seiler, 1973.
Study E = Geiger and Neal, 1981.
Study F = Gilbert et al., 1980.

Table 9 summarizes the mutagenic and related studies with test systems other than the Salmonella bacterial test system. This includes some recent data by Hay et al. (1983) presented at the recent ACS meeting in Washington, DC. Some of these *in vitro* studies used test systems such as E. coli, Saccharomyces sp., African Blood Lily, etc. The *in vivo* studies include clastogenic studies and also morphologic cell transportation studies which have been done on either human or animal cells. Overall, most *in vivo* studies have given negative results, and TCDD is considered to have a relatively low potential for mutagenesis.

Table 9. Mutagenic and related studies with TCDD in test systems other than salmonella bacteria.

Tests/Reference	Results
Prophage induction in E. coli K39 (Hussain et al., 1972)	+

Table 9. Continued.

Tests/Reference	Results
Streptomycin independence in E. coli Sd-4 (Hussain et al., 1972)	+
Saccharomyces cerevisiae yeast (Bronzetti et al., 1980)	+
QB Virus RNA transfectivity (Kondorosi et al., 1973)	-
Chromosomal analysis of African Blood Lily (Jackson, 1972)	+
In vitro baby hamster kidney cells (Hay et al., 1983)	+
In vitro mouse lymphoma cells (Rogers et al., 1982)	+
Dominant lethal study in rats (Khera and Ruddick, 1973)	-
Cytogenetics, rat bone marrow (Green and Moreland, 1975)	-
Cytogenetics, rat bone marrow (Green et al., 1977)	±
Morphology of human and animal cells in culture (Beatty et al., 1975)	-
Cytogenetics, human blood cells, Seveso, Italy (Tenchini, 1979) (Reggiani, 1980)	-

Table 10 lists additional data regarding the mechanism of TCDD carcinogenesis. These data are more supportive of a non-genetic mechanism of TCDD carcinogenesis rather than a genetic mechanism. The data of Poland and Glover (1979) indicate a relative lack of any significant covalent binding of TCDD to DNA (four to six orders of magnitude less than most carcinogens). Second, Matsumura (1983) has proposed a mechanism of action for dioxin based on the alterations of cellular membranes and the subsequent potential effect on cell-to-cell communication. A Swedish group of researchers, headed by Drs. Ahlborg and Thunberg, have published extensively on the role that Vitamin A may play in the mechanism of action of TCDD (Thunberg, 1983). From a pathologic assessment, some of the lesions of dioxin toxicity in animals have similarity to Vitamin A deficiency.

Another factor supporting a nongenetic mechanism of TCDD carcinogenesis is based on the observation of carcinogenic response in animal studies at dose levels that are actually producing frank toxicity (Kociba et al., 1978).

Table 10. Data supportive of a nongenetic rather than genetic mechanism of TCDD carcinogenesis.

-- Lack of covalent binding to DNA, with maximal binding 4-6 orders of magnitude less than most carcinogens (Poland and Glover, 1979).

-- Alteration of cellular membrane and effect on cell-to-cell communication (Matsumura, 1983).

-- Reduction of hepatic levels of Vitamin A as proposed mechanism of action (Thunberg, 1983).

-- Mixed response with *in vitro* mutagenic tests and lack of positive response with *in vivo* tests (Tables 8 and 9).

-- TCDD promotion of DEN-initiated liver tumors (target tissue) in rats (Pitot et al., 1980).

-- TCDD promotion of DMBA + MNNG initiated skin tumors (Poland et al., 1982).

-- Carcinogenic response in rodents occurring above dose levels producing organ toxicity (Kociba et al., 1978).

The study of TCDD toxicity and carcinogenicity in rodents that has been utilized in most risk calculations is the two-year study of TCDD in rats reported by Kociba et al. (1978). The data from that study, which addressed both chronic toxicity and carcinogenicity, have been used in several methods of extrapolation to man. The U.S. Food and Drug Administration (FDA) Bureau of Foods (Cordle, 1981) has utilized the no-effect-level defined in that study for the estimation of the margin of safety associated with the consumption of fish containing TCDD. Similarly, the Dutch Institute of National Health has since applied a safety factor in calculation of an acceptable daily intake (ADI) for man's exposure to TCDD (Van der Heijden et al., 1982). Conversely, the EPA Cancer Assessment Group has typically taken a nonthreshold approach and applied a linearized multi-stage extrapolation model to the study data in deriving a cancer-risk estimate. Thus, these data have been used in both threshold and nonthreshold types of extrapolations.

The two-year study of TCDD by Kociba et al. (1978) utilized the Sprague-Dawley rat with continuous ingestion of diets supplying TCDD at dose levels of 0.1, 0.01, 0.001 ug/kg/day. Groups of 50 rats of each sex were assigned to each of three treatment levels, and there were 86 controls of each sex. The dose-related toxicity that was noted for the groups given the 2 higher dose levels of 0.1 and 0.01 ug/kg/day has been described previously (Kociba et al., 1978). Table 11 lists the tumor and tumor-like lesions attributed to lifetime ingestion of TCDD by rats in that study. The data on hepatocellular lesions in female rats are the quantitative data upon which most of the risk extrapolations have been based.

Table 11. Increased incidence of tumor and tumor-like lesions attributed to ingestion by rats of diets containing TCDD for two years.

Tissue/Organ	Dose Level of TCDD (ug/kg/day)			
	0	0.1	0.01	0.001
Males				
Hard Palate/Nasal Turbinates				
Stratified squamous cell carcinoma	0/85	4/50*	0/50	0/50
Tongue				
Stratified squamous cell carcinoma	0/85	3/50*	1/50	1/50
Lung				
Squamous cell carcinoma	0/85	1/50	0/50	0/50
Females				
Hard Palate/Nasal Turbinates				
Stratified squamous cell carcinoma	0/86	4/49*	1/50	0/50
Tongue				
Stratified squamous cell carcinoma	1/86	2/49	0/50	0/50
Lung				
Squamous cell carcinoma	0/86	7/49*	0/50	0/50

Table 11. Continued.

Tissue/Organ	Dose Level of TCDD (ug/kg/day) 0	0.1	0.01	0.001
Females (continued)				
Liver				
Hepatocellular carcinoma	1/86	11/49*	2/50	0/50
Hepatocellular nodule(s)	8/86	23/49*	18/50*	3/50

*Statistically increased above control data, $p<0.05$.

Table excludes other data on incidence of tumors considered unrelated to treatment with TCDD.

(after Kociba et al., 1978)

Whereas the hepatic lesions can be assumed to have been mechanistically mediated via a nongenetic, systemic effect of TCDD, the stratified squamous cell carcinomas described for the hard palate, tongue or lung of rats of the high dose level was likely mediated via a nongenetic effect associated wih presumably localized higher concentrations or deposits of TCDD resulting from ingesting the TCDD as a dietary component of a finely pulverized rat chow. It is known that portions of pulverized rat chow, hair shafts and debris become embedded in various portions of the oral cavity and tongue and also may be aspirated into the rat lungs. It is possible that TCDD elicited this localized epithelial response that morphologically presented itself as a squamous cell carcinoma from the effect of higher levels of TCDD concentrating or deposited at these sites. The possibility is supported by a comparison of the results of Kociba et al. (1978), where TCDD was given in a pulverized diet, with the results of the NTP study where similar dose levels of TCDD were given to the same species but using oral intubation rather than dietary administration. In that oral intubation study where presumably lower levels of TCDD would be concentrated or deposited in the oropharynx and lungs, TCDD induced the liver tumors but did not induce the squamous cell carcinomas as was observed in the oropharynx and lungs of rats at the highest dose level of treatment in the study by Kociba et al. (1978).

Table 12 lists the incidence data for those tumor types for which a decrease was attributed to TCDD treatment. The tumor types decreased by TCDD treatment were those that typically occur at high spontaneous incidence rates in this particular Sprague-Dawley strain of rat. This included neoplasms of the uterus, mammary glands, and pituitary gland. It is difficult to ascertain whether these decreased incidences were a secondary reflection of the lack of obesity, etc., that can be associated with a decreased tumor incidence, or were a bona fide anticarcinogenic effect as described by Cohen et al. (1979) for TCDD.

Table 12. Decreased incidence of tumors attributed to ingestion by rats of diets containing TCDD for two years.

Tissue/Organ	Dose Level of TCDD (ug/kg/day) 0	0.1
Males		
Pancreas	14/25	2/50*
Acinar adenoma	28/85	4/50*
Females		
Uterus		
Benign tumor	30/86	8/49*
Mammary gland		
Benign tumor	73/86	24/49*
Pituitary		
Adenoma	43/86	12/49*

*Statistically decreased from control data, $p<0.05$.

Data at lower dose levels comparable to controls, except for possible decrease in incidence of benign uterine tumors in females given 0.01 ug/kg/day.

(after Kociba et al., 1978)

Table 13 shows a correlation of dosage, biologic response, and tissue levels of TCDD at termination of the two-year study. The dose levels of 0.1, 0.01, and 0.001 ug/kg/day equated with approximately 2200, 210, and 22 parts per trillion of TCDD in the diet, respectively. The correct way to utilize these data is to base calculations upon the dosage in ug/kg body weight/day rather than parts per trillion (ppt) in the diet. Both carcinogenicity and other toxicity were observed at the highest dose level. At the intermediate dose level, some toxicity was observed but there was no definite carcinogenic response, with the maximal response represented by the hyperplastic nodules in the liver of the female rats. At the lower dose level of 0.001 ug/kg/day at which no carcinogenicity or other toxicity occurred, fat and liver tissue both contained 540 ppt of TCDD after the two years of continuous treatment. The terminal tissue levels of TCDD increased with increasing dose, as listed in Table 13.

Table 13. Correlation of dosage, response, and tissue levels of TCDD at termination of two-year toxicity study in rats.

Dose Levels of TCDD		Response		Terminal TCDD Content ppt	
ug/kg/day	ppt in Diet	Carcino-genicity	Other Toxicity	Fat	Liver
0.1	2200	Yes	Yes	8100	24000
0.01	210	No*	Yes	1700	5100
0.001	22	No	No	540	540

*Maximal response of hyperplastic nodules of liver in female rats.

(after Kociba et al., 1978)

This concludes my summary presentation of the experimental rodent data on the carcinogenicity of TCDD. Hopefully, it has laid the groundwork for the subsequent presentations on the quantitative risk assessment extrapolations that will follow.

REFERENCES

Beatty, P.W., Lembach, K.J., Holscher, M.A. and Neal, R.A. Effects of 2,3,7,8-Tetrachlorodibenzo-p-Dioxin (TCDD) on Mammalian Cells in Tissue Cultures. Toxicology and Applied Pharmacology 31:309-312, 1975.

Berry, D.L., DiGiovanni, J., Juchau, M.R., Bracken, W.M., Gleason, G.L. and Slaga, T.J. Lack of Tumor-Promoting Ability of Certain Environmental Chemicals in a Two-Stage Mouse Skin Tumorigenesis Assay. Research Communications in Chemical Pathology and Pharmacology 20(1):101-108, 1978.

Bronzetti, G., Lee, I., Zeiger, E., Malling, H. and Suzuki. Genetic Effects of TCDD In Vitro and In Vivo Using D7 Strain of S. Cerevisiae. Mutation Research 74:206-207, 1980.

Cohen, G.M., Bracken, W.M., Iyer, R.P., Berry, D.L., Selkirk, J.K. and Slaga, T.J. Anticarcinogenic Effects of 2,3,7,8-Tetrachlorodibenzo-p-dioxin on Benzo(a)pyrene and 7,12-Dimethylbenz(a)anthracene Tumor Initiation and Its Relationship to DNA Binding. Cancer Research 39:4027-4033, 1979.

Cordle, F. The Use of Epidemiology in the Regulation of Dioxins in the Food Supply. Regulatory Toxicology and Pharmacology 1:379-387, 1981.

DiGiovanni, J., Viaje, A., Berry, D.L., Slaga, T.J. and Juchau, M.R. Tumor-Initiating Ability of 2,3,7,8-Tetrachlorodibenzo-p-Dioxin (TCDD) and Arochlor 1254 in the Two-Stage System of Mouse Skin Carcinogenesis. Bulletin of Environmental Contamination and Toxicology 18(5):552-557, 1977.

Geiger, L.E. and Neal, R.A. Mutagenicity Testing of 2,3,7,8-Tetrachlorodibenzo-p-dioxin in Histidine Auxotrophs of Salmonella typhimurium. Toxicology and Applied Pharmacology 59:125-129, 1981.

Gilbert, P., Saint-Ruf, G., Poncelet, F. and Mercier, M. Genetic Effects of Chlorinated Anilines and Azobenzenes on Salmonella typhimurium. Archives of Environmental Contamination and Toxicology 9:533-541, 1980.

Grasso, P. and Goldberg, L. Review Section: Subcutaneous Sarcoma as an Index of Carcinogenic Potency. Food and Cosmetics Toxicology 4:297-320, 1966.

Green, S. and Moreland, F.S. Cytogenetic Evaluation of Several Dioxins in the Rat. Toxicology and Applied Pharmacology 33:161, 1975.

Green, S., Moreland, F. and Sheu, C. Cytogenetic Effect of 2,3,7,8-Tetrachlorodibenzo-p-Dioxin on Rat Bone Marrow Cells. FDA By-Lines Issue No. 6, May, 1977.

Hay, A., Ashby, J., Styles, J.A. and Elliott, B. Mutagenic Properties of 2,3,7,8-Tetrachlorodibenzo-p-Dioxin. Presentation to Symposium on Chlorinated Dioxins and Dibenzofurans in the Total Environment, II. American Chemical Society Meeting, Washington DC, August 29, 1983.

Hussain, S., Ehrenberg, L., Lofroth, G. and Gejvall, T. Mutagenic Effects of TCDD on Bacterial Systems. Ambio 1:32-33, 1972.

Jackson, W.T. Regulations of Mitosis III. Cytological Effects of 2,4,5-Trichlorophenoxyacetic Acid and of Dioxin Contaminants in 2,4,5-T Formulations. Journal of Cell Science 10:15-25, 1972.

Khera, K.S. and Ruddick, J.A. Polychlorodibenzo-p-dioxins: Perinatal Effects and the Dominant Lethal Test in Wistar Rats. In: Chlorodioxins - Origin and Fate, Etcyl H. Blair, Editor. Advances in Chemistry 120:70-84, 1973.

Kociba, R.J., Keyes, D.G., Beyer, J.E., Carreon, R.M., Wade, C.E., Dittenber, D.A., Kalnins, R.P., Frauson, L.E., Park, C.N., Barnard, S.D., Hummel, R.A. and Humiston, C.G. Results of a Two-Year Chronic Toxicity and Oncogenicity Study of 2,3,7,8-Tetrachlorodibenzo-p-Dioxin in Rats. Toxicology and Applied Pharmacology 46:279-303, 1978.

Kondorosi, A., Fedorcsak, I., Solymosy, F., Ehrenberg, L. and Osterman-Golkar, S. Inactivation of QB RNA by Electrophiles. Mutation Research 17:149-161, 1973.

Kouri, R.E., Rude, T.H., Jogelkar, R., Dansette, P.M., Jerina, D.M., Atlas, S.A., Owens, I.S. and Nebert, D.W. 2,3,7,8-Tetrachlorodibenzo-p-dioxin as Cocarcinogen Causing 3-Methylcholanthrene-initiated Subcutaneous Tumors in Mice Genetically "Nonresponsive" at Ah Locus. Cancer Research 38:2777-2783, 1978.

Matsumura, F. Biochemical Aspects of Action Mechanisms of 2,3,7,8-Tetrachlorodibenzo-p-Dioxin (TCDD) and Related Chemicals in Animals. Pharmacology and Therapeutics 19:195-209, 1983.

McCann, J. Unpublished data cited by Wassom et al., 1977.

National Toxicology Program (NCI). Bioassay of Dibenzo-p-Dioxin for Possible Carcinogenicity. NCI Technical Report No. 122, 1979a.

National Toxicology Program (NCI). Bioassay of 2,7-Dichlorodibenzo-p-Dioxin for Possible Carcinogenicity. NCI Technical Report No. 123, 1979b.

National Toxicology Program (NCI). Bioassay of a Mixture of 1,2,3,6,7,8-Hexachlorodibenzo-p-Dioxin for Possible Carcinogenicity. NTP Report No. 80-12, 1980. NCI Report No. 198, 1980.

National Toxicology Program. Carcinogenesis Bioassay of 2,3,7,8 Tetrachlorodibenzo-p-Dioxin in Osborne-Mendel Rats and B6C3Fl Mice (gavage study). NTP Technical Report No. 209, 1982. NIH Report No. 82-1765, 1982a.

National Toxicology Program. Carcinogenesis Bioassay of 2,3,7,8-Tetrachlorodibenzo-p-Dioxin in Swiss-Webster Mice (dermal study). NTP Technical Report No. 201, NIH Report No. 82-1757, 1982b.

Nebert, D.W., Thorgeirsson, S.S. and Felton, J.S. "Genetic Differences in Mutagenesis, Carcinogenesis and Drug Toxicity", in In Vitro Metabolic Activation in Mutagenesis Testing (F.J. de Serres, J.R. Fouts, J.R. Bend, R.M. Philpot, eds.), Elsevier/North-Holland Biomedical Press, Amsterdam, 105, 1976.

Pitot, H.C., Goldsworthy T., Campbell, H.A. and Poland, A. Quantitative Evaluation of the Promotion by 2,3,7,8-Tetrachlorodibenzo-p-dioxin of Hepatocarcinogenesis from Diethylnitrosamine. Cancer Research 40:3616-3620, 1980.

Poland, A., Palen, D. and Glover, E. Tumour promotion by TCDD in skin of HRS/J hairless mice. Nature 300:271-273, 1982.

Poland, A. and Glover E. An Estimate of the Maximum in Vivo Covalent Binding of 2,3,7,8-Tetrachlorodibenzo-p-dioxin to Rat Liver Protein, Ribosomal RNA, and DNA. Cancer Research 39:3341-3344, 1979.

Reggiani, G. Acute Human Exposure to TCDD in Seveso, Italy. Journal of Toxicology and Environmental Health 6:27-43, 1980.

Rogers, A.M., Andersen, M.E. and Back, K.C. Mutagenicity of 2,3,7,8-Tetrachlorodibenzo-p-dioxin and Perfluoro-n-decanoic Acid in L5178Y Mouse-lymphoma Cells. Mutation Research 105:445-449, 1982.

Seiler, J.P. A Survey on the Mutagenicity of Various Pesticides. Experientia 29:622-623, 1973.

Squire, R. An Assessment of the Experimental Evidence for Potential Carcinogenicity of Hexachlorodibenzo-p-dioxins. Robert S. Squire Associates, Inc., Ruxton, MD, 1983.

Tenchini, M.L., Crimaudo, C., Simoni, G., DeCarli, L., Giorgi, R. and Nuzzo, F. "Approaches to the Evaluation of Genetic Damage After a Major Hazard in Chemical Industry: Preliminary Cytogenetic Findings on TCDD-Exposed Subjects After the Seveso Accident", in Genetic Damage in Man Caused by Environmental Agents (K. Berg, ed.), Academic Press, New York, 301-317, 1979.

Toth, K., Somfai-Relle, S., Sugar, J. and Bence, J. Carcinogenicity Testing of Herbicide 2,4,5-Trichlorophenoxyethanol Containing Dioxin and of Pure Dioxin in Swiss Mice. Nature 27:548-549, 1979.

Thunberg, T. Studies on the Effect of 2,3,7,8-Tetrachlorodibenzo-p-dioxin on Vitamin A: A New Aspect Concerning the Mechanism of Toxicity. Publication of Karolinska Institute, Stockholm, Sweden, 1983.

Van der Heijden, C.A., Knaap, A.G., Kramers, P.G. and Van Logten, M.J. Evaluation of the Carcinogenicity and Mutagenicity of 2,3,7,8-Tetrachlorodibenzo-1,4-dioxin (TCDD); Classification and No-effect-level. State Institute of National Health, Bilthoven, Netherlands, 1982.

Wassom, J.S., Huff, J.E. and Loprieno, N. A Review of the Genetic Toxicology of Chlorinated Dibenzo-p-Dioxins. Mutation Research 47:141-160, 1977/1978.

STATISTICAL ANALYSIS OF THE CARCINOGENESIS BIOASSAY DATA RELATING TO THE RISKS FROM EXPOSURE TO 2,3,7,8-TETRACHLORODIBENZO-*p*-DIOXIN

Christopher J. Portier, Ph.D.*
David G. Hoel, Ph.D.*
John Van Ryzin, Ph.D.**

*Biometry and Risk Assessment Program
National Institute of Environmental Health Sciences
Research Triangle Park, North Carolina 27709

**Division of Biostatistics
Columbia University
New York, New York 10032

From Public Health Risks of the Dioxins, proceedings of a symposium held on October 19-20, 1983 at The Rockefeller University, New York City. Edited by William W. Lowrance. Published by William Kaufmann, Los Altos, California, 1984.

INTRODUCTION

Much attention has been focused recently on the quantitative estimation of risk due to exposure from chemical and physical agents released into the environment (NAS, 1983). Of the toxic endpoints related to exposure, the endpoint which has received the most attention is cancer. The National Academy of Sciences has defined four steps in the risk assessment process: (1) hazard identification, (2) dose-response assessment, (3) exposure assessment, and (4) risk characterization. In this paper we will restrict ourselves to the second step, dose-response assessment. The other steps are discussed elsewhere in this symposium. We will focus primarily in this report on the dose-response assessment in the experimental animal and briefly comment on the problem of extrapolation of the animal results to man.

Recent events in Italy and the United States have increased public concern over exposure to 2,3,7,8-tetrachlorodibenzo-p-dioxin (2,3,7,8-TCDD). There are 75 possible polychlorinated dibenzo-p-dioxins (PCDD) of which the 2,3,7,8 isomer is considered the most toxic (McConnell, Moore, Haseman, and Harris, 1978). PCDDs are the by-products of certain manufacturing processes. The process of major public interest is the manufacture of 2,4,5-trichlorophenol (2,4,5-T). In the course of producing 2,4,5-T, 2,3,7,8-TCDD is formed. Much of the current information on human exposure to 2,3,7,8-TCDD is due to accidents in the manufacture of 2,4,5-T (Hay, 1982). In addition, many people were potentially exposed to 2,3,7,8-TCDD in Agent Orange, a herbicide mixture containing 50% 2,4,5-T which was used for defoliation during the Vietnam War (NAS, 1974). Another route of exposure to 2,3,7,8-TCDD which is of general public concern is waste disposal sites such as Love Canal (Axelrod, 1980) and general waste such as occurred in Times Beach, Missouri (Kimbrough, Carter, Liddle, Cline and Phillips, 1977).

In this paper, we will review some of the available animal data on the cancer risk associated with exposure to 2,3,7,8-TCDD. Statistical procedures for "low dose extrapolation" using the multistage model will be applied to these data and estimates of risk at various exposure levels will be given. In addition, it is common in studies of this type to group certain or all cancers and sites and analyze this grouped response. We will discuss the interpretation and correctness of this approach.

ANIMAL BIOASSAYS

Consider a laboratory cancer bioassay in which N animals are exposed to differing levels of some possibly carcinogenic agent. Let d_i denote the differing dose levels and let n_i denote the number of animals started on study at each of these doses, $i=0,1,...k$. Let x_i denote the number of animals receiving dose d_i which have been observed to have the cancer(s) of interest by study termination. Define $p_i=x_i/n_i$. One method of estimating the low-dose effects associated with this agent would be to fit (via maximum likelihood) some dose-response model, P(d), through the (d_i,p_i) pairs, with the n_i used as weights.

Many models exist for estimating the dose-response curve. A difficulty with estimating low-dose risks from these models is that a number of these models will fit the observed data equally well so that it is mathematically impossible to distinguish between them based on goodness-of-fit, but they will yield markedly different low-dose risk estimates (Chand and Hoel, 1974). Thus the form of the dose-response model should be based on biological theory. A quantal response model which has some basis in biological theory is the multistage model (Armitage and Doll, 1954) which can be expressed mathematically (Crump, Hoel, Langley, and Peto, 1976) as

$$P(d)=1-\exp\left(-\sum_{j=0}^{L} a_j d^j\right) \qquad (1)$$

where $a_j \geq 0$ for all j, $j=0,1...L$ and L fixed.

The number of parameters which can fit for a particular set of data will depend upon the number of dose groups (including control) used in the bioassay. Although there are methods which allow for higher order polynomial fits, we will restrict the fitted model to have order no greater than the number of dose groups minus one. Thus in a three dose and control experiment, we will fit model parameters a_0, a_1, a_2, and a_3. The best fitting model may set some of these parameters to zero.

In many studies, the treatment will affect the survival among the various dose groups. In these cases, it is appropriate to use one of the various time-to-tumor models. These time-to-tumor models specify the probability of dying with the tumor by time t given a dose of d. We will denote this as P(t;d). Note that we are assuming the tumor-of-interest is incidental in nature; that is, the animals typically will die of something other than the tumor-of-interest and the tumor-of-interest is simply an incidental finding. For a complete discussion of how these models are fit see Kalbfleish, Krewski and Van Ryzin (1983) and for a discussion of how incidental and lethal tumors affect the risk assessment see Portier and Hoel (1984). We will assume the product form of the hazard in the time-to-tumor model (Kalbfleish and Prentice, 1980) again using the multistage form for dose. Using a Weibull form for the time component of the model, we have the multistage-Weibull model which can be expressed mathematically as

$$P(t;d)=1-\exp(-(\sum_{j=0}^{L} a_j d^j)\ (t-w)^r) \qquad (2)$$

where $a_j \geqslant 0$, $w \geqslant 0$ and $r>0$. The restrictions on the number of dose parameters applied to model (1) will also be applicable to model (2).

There are several measures which could be used to determine the potential risk associated with exposure to some toxic substance (see e.g., Hoel, 1982). We will restrict this analysis to the more traditional measure, the virtually safe dose (VSD). For quantal response models, the VSD is that dose which will cause a small ε increase in response over background in a lifetime study.

Mathematically, the VSD is defined as that dose which satisfies the equation

$$\varepsilon = P(VSD)-P(0).$$

For the time-to-tumor models that assume tumors occur incidentally, the VSD is that dose which, given a prespecified time, will cause a small (ε) increase in the prevalence of the tumor. Mathematically, the VSD satisfies

$$\varepsilon = P(t;VSD)-P(t;0)$$

for t fixed.

We will not list or review all of the animal bioassay data relevant to the chemical 2,3,7,8-TCDD. There are several agency risk assessments of 2,3,7,8-TCDD which provide excellent reviews of this literature (Albert, 1980; Kimbrough, Falk, Stehr and Fries, 1983). We will concentrate on the two carcinogenesis bioassays which seem to have few flaws and which best quantify the dose-response relationship. The first study was performed by Dow Chemical Company (Kociba et al., 1978). The second study was done by the National Toxicology Program (NTP, 1982).

The Dow study was a feeding study done on male and female Sprague-Dawley rats, Spartan substrain. There were 86 untreated control animals for each sex and 50 animals per sex exposed to each of three dose levels: 0.001, 0.01 and 0.1 micrograms per kilogram body weight per day (ug/kg.bw/day). A substantial proportion of the animals in each group, including the control group, died before the two-year sacrifice (78% to 92% in the males and 68% to 92% in the females). The important lesion sites for risk assessment in the two sexes are given in Table 1 along with the incidence counts at each dose. The original pathology done by Dow was reviewed by Squire (Albert, 1980). These counts are also given in Table 1.

Table 1. Important lesion sites in the Dow TCDD study.

Site	Sex	Pathologist	Control	0.001	0.01	0.1
Stratified squamous cell carcinoma of the tongue	Males	Kociba	0/76*	1/49	1/49	3/42
		Squire	0/77	1/44	1/49	3/44
Nasal turbinates/ hard palate squamous cell carcinoma	Males	Kociba	0/51	0/34	0/27	4/30
		Squire	0/55	1/34	0/26	6/30
	Females	Kociba	0/54	0/30	1/27	4/24
		Squire	0/54	0/30	1/27	5/22
Hepatocellular nodules and carcinoma	Females	Kociba	9/86	3/50	18/50	34/48
		Squire	16/86	8/50	27/50	33/47
Lung keratinizing squamous cell carcinoma	Females	Kociba	0/86	0/50	0/49	7/49
		Squire	0/86	0/50	0/49	8/47

*Entries are number of animals with the tumor/number of animals examined at that site.

The NTP study was a gavage experiment involving both sexes of $B6C3F_1$ mice and Osborne-Mendel rats. There were 75 vehicle (corn oil-acetone) treated control animals and 50 animals at each of three doses for each sex by species combination. The doses were administered twice weekly in ug/kg.bw. In order to use the same scale, the total weekly dose has been divided by seven yielding average daily doses (this assumes that twice weekly peaks in 2,3,7,8-TCDD body burden are equivalent to average daily doses). With this conversion, both sexes in rats and male mice received doses of 0.0014, 0.0071, and 0.0714 ug/kg.bw/day. Female mice received doses of 0.0057, 0.0286, and 0.2859 ug/kg.bw/day. There were no significant survival differences among groups. The sites/ morphologies that had qualitative increases are given in Table 2 along with the quantal response.

Table 2. Important lesion sites in the NTP bioassay of 2,3,7,8-TCDD.

Site	Sex/Species	Control	Low	Medium	High
Thyroid Follicular-Cell Adenoma	Male Rat	1/69*	5/48	6/50	10/50
Thyroid Follicular-Cell Adenoma	Female Rat	3/73	2/45	1/49	6/47
Liver Neoplastic Nodules	Female Rat	5/75	1/49	3/50	12/49
Liver Neoplastic Nodules or Carcinoma	Female Mouse	5/75	1/49	3/50	14/49
Adrenal Cortical Adenoma or Carcinoma	Female Rat	11/73	9/49	5/49	14/46
Liver Hepatocellular Carcinoma	Male Mouse	8/73	9/49	8/49	17/50
Liver Carcinoma and Adenoma	Male Mouse	15/73	12/49	13/49	27/50
Hepatocellular Carcinoma	Female Mouse	1/73	2/50	2/48	6/47
Hepatocellular Carcinoma or Adenoma	Female Mouse	3/73	6/50	6/48	11/47
Thyroid Follicular-Cell Adenoma	Female Mouse	0/69	3/50	1/47	5/46
Subcutaneous Tissue Fibrosarcoma	Female Mouse	1/74	1/50	1/48	5/47
Lymphoma or Leukemia	Female Mouse	18/74	12/50	13/48	20/47

* Entries are number of animals with the tumor/number of animals examined at that site.

Table 3 gives the VSD estimates for one-in-ten-thousand added risk (1.E-4) and one-in-one-million (1.E-6) added risk and respective lower confidence bounds using the multistage model (1) for each positive tumor site for each sex in the Kociba study.

Table 3. Estimates and approximate 95% lower confidence bounds (LCB) for the VSD[1] of TCDD from the Dow study in Sprague-Dawley rats.

Lesion/Pathologist	Model[2]	Added Risk 1.E-4		Added Risk 1.E-6		Chi-squared G-O-F[4]
Male rat stratified squamous cell carcinoma of the tongue						
Kociba	Linear*	1.4E+5	(5.9E+4)	1.4E+3	(6.0E+2)	1.43
Squire	Linear*	1.5E+5	(6.3E+4)	1.5E+3	(6.3E+2)	1.60
Male rat nasal turbinates or hard palate squamous cell carcinoma						
Kociba	Cubic*	8.9E+6	(1.7E+4)	1.9E+6	(1.8E+2)	0.00
	Linear	7.8E+4	(3.8E+4)	7.8E+2	(3.8E+2)	0.39
Squire	Cubic*	7.8E+6	(1.2E+4)	1.7E+6	(1.2E+2)	2.39
	Linear	4.9E+4	(2.6E+4)	4.9E+2	(2.6E+2)	2.72
Female rat lung keratinizing squamous cell carcinoma						
Kociba	Cubic*	8.7E+6	(3.2E+4)	1.9E+6	(3.3E+2)	0.01
	Linear	7.3E+4	(4.1E+4)	7.3E+2	(4.1E+2)	0.82
Squire	Cubic*	8.1E+6	(2.7E+4)	1.8E+6	(2.7E+2)	0.01
	Linear	6.0E+4	(3.5E+4)	6.0E+2	(3.6E+2)	0.98
Female rat nasal turbinates or hard palate squamous cell carcinoma						
Kociba	Linear*	5.0E+4	(2.6E+4)	5.0E+2	(2.6E+2)	0.49
Squire	Linear*	3.7E+4	(2.0E+4)	3.7E+2	(2.0E+2)	0.20

Table 3. Continued.

Lesion/Pathologist	Model[2]	Added Risk 1.E-4		1.E-6		Chi-squared G-O-F[4]
Female rat liver hepatocellular carcinoma or adenoma						
Kociba	Linear*	7.7E+3	(5.7E+3)	7.7E+1	(5.7E+1)	6.35
	Transf[3]	3.8E+3	(2.9E+3)	3.8E+1	(2.9E+1)	2.18
Squire	Linear*	8.6E+3	(6.1E+3)	8.6E+1	(6.1E+1)	10.36
	Transf[3]	3.8E+3	(2.8E+3)	3.8E+1	(2.8E+1)	4.30

[1]Entries are VSD (LCB) in femtograms/kg body weight/day using the multistage model. Notation: $XE + b = x \cdot 10^b$

[2]Linear Model: $P(d)=1-\exp(-a_0-a_1d)$
Cubic Model: $P(d)=1-\exp(-a_0-a_3d^3)$

[3]A multistage model is fit to the average liver concentration of TCDD and safe doses are transformed by using the relationship

administered dose = liver dose/510.297

The divisor is the least-squares slope between the administered and the average liver dose of TCDD for control and the two lowest doses.

[4]G-O-F = goodness of fit.

*best fit.

The lower confidence bounds were produced by the program GLOBAL79 (Crump and Watson, 1979). Under "G-O-F" in Table 3, we give the simple Pearson chi-squared statistic for determining goodness-of-fit. Note that since we fit as many model parameters as there are data points, this value technically has zero degrees of freedom so a formal test would be of little value. However, the value is still a meaningful guide for determining deviation of the fitted model from the data under the constraints which were imposed. Which parameters of the dose polynomial are non-zero is given under the column labelled "Model". As fitted, the linear model has $a_0>0$, $a_1>0$, $a_2=0$ and $a_3=0$; the cubic model has $a_0>0$, $a_1=0$, $a_2=0$ and $a_3>0$. In addition, when the best fit to the data was not a linear model, the linear model was fit to the data to provide an upper bound on the possible models which may fit the data and a lower bound on the possible VSD.

Except for hepatocellular tumors in the females, the multistage model seems to adequately fit the data. In the case of female liver tumors, the best fit was the linear model and the chi-squared values were 6.35 for the Kociba pathology report and 10.36 for Squire's interpretation. The reason for this lack of fit is due to a plateau in response beyond a dose of 0.01 ug/kg.bw/day. This is possibly due to some type of non-linear absorption of the isomer (for a discussion of the non-linear kinetic models, see Hoel, Kaplan and Anderson, 1983; Van Ryzin and Rai, 1983). In the original manuscript, Dow gives the average concentration of 2,3,7,8-TCDD in the liver based upon a random sample of five rats from each group at terminal sacrifice. We assume this concentration held throughout the major portion of each animal's lifetime. For females, these concentrations are 0.540 parts per billion (ppb), 5.1 ppb and 24 ppb for the daily doses 0.001 ug/kg.bw, 0.01 ug/kg.bw, and 0.1 ug/kg.bw respectively. If we fit the response to ppb concentration of 2,3,7,8-TCDD in the liver instead of the ug/kg.bw/day dose, the resulting multistage model for Squire's pathology is

$$P(d') = 1-\exp(-0.257-0.0521\ d')$$

where d' is measured in ppb 2,3,7,8-TCDD in the liver. The chi-squared value for this model is 4.3, an improvement over the original 10.36. Similar results hold for the Kociba pathology review.

Thus far, every VSD we have given has been expressed in ug/kg.bw/day. To remain consistent, we would like to transform the results above from ppb in the liver to ug/kg.bw/day. If we assume a zero dose would result in a 0 ppb concentration in the liver, then the three pairs (0, 0), (0.54, 0.001) and (5.1, 0.01) nearly form a straight line (see Figure 1).

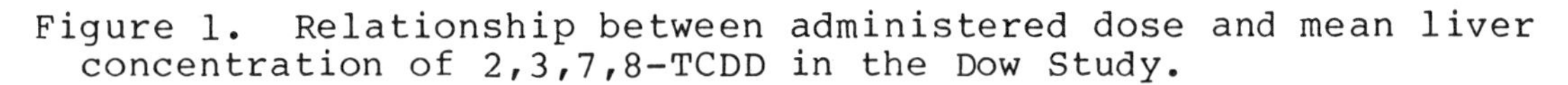

Figure 1. Relationship between administered dose and mean liver concentration of 2,3,7,8-TCDD in the Dow Study.

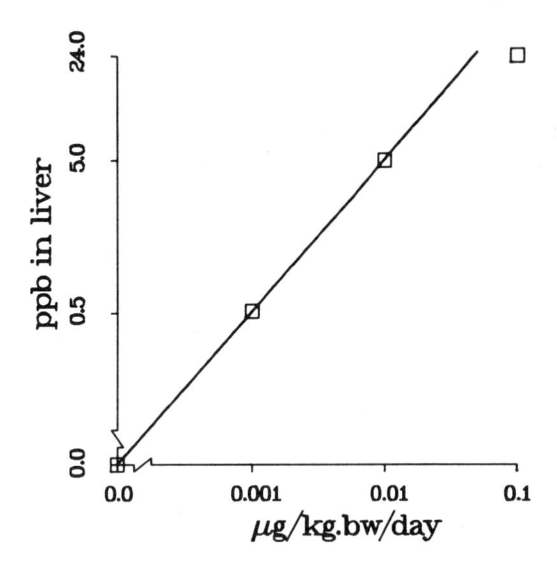

Since any VSD estimate would be below 0.001 ug/kg.bw/day, we can use this linear relationship to convert from liver concentration to administered dose. This relationship is:

dose (in ug/kg.bw/day) = dose (in ppb in the liver)/510.

These transformed ("Transf") VSD estimates and lower confidence bounds are given in Table 3 along with the chi-squared goodness-of-fit statistics.

We were able to obtain from Dow the individual animal survival times expressed in multi-week intervals and the individual death times for female rats dying with a liver tumor. To fit these data to the multistage-Weibull model, animals that died within an interval were assumed to have died at the midpoint of the interval. The multistage-Weibull model was fit to these data and the resulting estimate of the VSD at 729 days (the length of the study) was 18.09 femtograms for an added risk of one-in-a-million. This is approximately one-fourth of what was estimated using the simple multistage model and one-half of what was obtained using the transformed data. Based on graphical techniques outlined by Brown and Hoel (1983), this model seems to fit the data adequately.

It is important to understand the difference between VSDs expressed from the quantal response models and those from the time-to-tumor model. The quantal response model predicts the lifetime tumor incidence in a given population. Thus, if one million rats are exposed to the VSD for an added risk of 1.E-6 and followed for their entire lifetime (or the length of this study, 729 days), one would expect to see one additional animal with a tumor over the number expected in the absence of any exposure. On the other hand, the time-to-tumor model predicts the prevalence of the tumor in the population at a given time. Thus, if one million rats lived to be 729 days old and if all of these animals had received the VSD calculated from the time-to-tumor model and if every animal were sacrificed at 729 days, one would expect one additional animal with a tumor. Unless every animal on study survives to 729 days, the initial population would have to be larger than one million animals. Thus, in comparing the VSD from a quantal response model with the VSD from a time-to-tumor model, it may make little sense to use the same added risk. In addition, when the time-to-tumor model is used, it may seem reasonable to calculate several VSD estimates for different times and different added risks.

For the NTP study, the estimates of the VSD from model (1) and the associated chi-squared goodness-of-fit statistics are given in Table 4.

Table 4. Estimates and approximate 95% lower confidence bounds (LCB) for the VSD[1] of TCDD from the NTP study.

Lesions	Model[2]	Added Risk 1.E-4		Added Risk 1.E-6		Chi-Squared G-O-F[3]
Male rat thyroid						
Follicular-cell adenoma	Linear*	4.0E+4	(2.1E+4)	4.0E+2	(2.1E+2)	4.81
Female rat thyroid						
Follicular-cell adenoma	Cubic*	7.1E+6	(3.7E+4)	1.5E+6	(3.7E+2)	0.48
	Linear	7.6E+4	(3.5E+4)	7.6E+2	(3.5E+2)	0.79

Table 4. Continued.

Lesions	Model[2]	Added Risk 1.E-4	1.E-6	Chi-Squared G-O-F[3]
Female rat liver				
Neoplastic nodules	Lin-Qua*	1.4E+6 (1.9E+4)	1.1E+5 (1.9E+2)	1.31
	Linear	3.2E+4 (1.9E+4)	3.2E+2 (1.9E+2)	1.61
Nodules and carcinomas	Quad*	1.3E+6 (1.7E+4)	1.3E+5 (1.7E+2)	1.31
	Linear	2.6E+4 (1.6E+4)	2.6E+2 (1.6E+2)	1.74
Male mouse liver				
Hepatocellular carc.	Linear*	2.6E+4 (1.5E+4)	2.6E+2 (1.5E+2)	1.01
Adenomas and carc.	Linear*	1.3E+4 (8.6E+3)	1.3E+2 (8.6E+1)	0.14
Female mouse liver				
Hepatocellular carc.	Linear*	2.5E+5 (1.2E+5)	2.5E+3 (1.2E+3)	0.74
Adenomas and carc.	Linear*	1.5E+5 (7.8E+4)	1.4E+3 (7.8E+2)	2.59
Female mouse thyroid				
Follicular-cell adenoma	Linear*	3.0E+5 (1.4E+5)	3.0E+3 (1.4E+3)	3.75
Female mouse sarcoma subcutaneous tissue	Lin-Cub*	4.3E+5 (1.4E+5)	4.3E+3 (1.4E+3)	0.05
	Linear	3.0E+5 (1.4E+5)	3.0E+3 (1.4E+3)	0.06
Female mouse lymphoma and leukemia	Linear*	1.0E+5 (5.4E+4)	1.0E+3 (5.4E+2)	0.02

[1]Entries are VSD (LCB) in femtograms/kg body weight/day using the multistage model. Notation: XE + b = $x \cdot 10^b$

[2]Linear: $P(d)=1-\exp(-a_0-a_1d)$ Quad: $P(d)=1-\exp(-a_0-a_2d^2)$

Cubic: $P(d)=1-\exp(-a_0-a_3d^3)$ Lin-Qua: $P(d)=1-\exp(-a_0a_1d-a_2d^2)$

Lin-Cub: $P(d)=1-\exp(-a_0-a_1d-a_3d^3)$

[3]G-O-F = goodness of fit.

*best fit.

As in the analysis of the Kociba study, when the linear model was not the best fit, it was fit separately. For the multistage model and in those cases where we specifically fit the linear model, the fit was adequate for all sites/sex/species.

Because survival was not related to dose and most of the animals survived to sacrifice, the multistage-Weibull model gave results that differed by no more than 5% from the dichotomous response results given in Table 4. In all cases, the VSD estimates from the multistage model (1) were slightly larger than those for the multistage-Weibull model (which follows from the discussion given earlier).

In the Dow study, the lowest estimated VSD in male rats was for stratified squamous cell carcinoma of the tongue. For a one-in-a-million added risk, this was 1426 femtograms (fg)/kg.bw/day using the Kociba pathology results and 1518 fg/kg.bw/day using the Squire pathology results. A femtogram is 10^{-15} grams. If we limit the discussion to only linear models, then the smallest VSD estimate for male rats was for nasal turbinates or hard palate squamous cell carcinomas. For one-in-a-million added risk, the VSD estimates were 775 fg/kg.bw/day and 495 fg/kg.bw/day for the Kociba and Squire reviews respectively. The smallest 95% lower confidence bound on the VSD was 117 fg/kg.bw/day.

For female rats in the Kociba study, the lowest estimates of the VSD were for liver tumors (adenomas or carcinomas). Using the transformed model, the VSD estimate for one-in-a-million added risk was 38 fg/kg.bw/day and a 95% lower confidence bound of 29.1 fg/kg.bw/day for the Kociba review and 27.6 fg/kg.bw/day for the Squire review. These numbers dropped by a factor of approximately 2 when a time-to-tumor analysis using the multistage-Weibull model was fitted.

In the NTP study, the lowest estimated VSDs were as follows: male rat thyroid follicular cell adenoma, 403 fg/kg.bw/day; female rat liver nodules or carcinomas, 256 fg/kg.bw/day; male mouse liver adenomas and carcinomas, 134 fg/kg.bw/day; female mouse leukemia/lymphoma, 1028 fg/kg.bw/day. The lower confidence bounds on the VSD at these sites were 214 fg/kg.bw/day, 160 fg/kg.bw/day, 86 fg/kg.bw/day and 543 fg/kg.bw/day respectively.

Of the six sex/species groups studied, 2,3,7,8-TCDD appears to cause an increase in liver tumors in four of the sex/species groups and in at least one sex of all three species. In addition, in three of the four liver affected sex/species groups, the liver was the most sensitive site judging by the estimates of the VSD.

DISCUSSION

We have presented one possible analysis of the bioassay results on 2,3,7,8-TCDD. As mentioned previously, other models and approaches could have been used. In some cases, these methods would have resulted in much larger estimates of virtually safe doses than we have given (Krewski et al., 1983). The method we have presented has a plausible biological basis and is the one which would typically yield the largest estimates of risk.

In other analyses of these bioassays and for other chemicals it is common to group very distinct tissues together and look at either all tumor bearing animals or collectively to look at all tumor sites that showed a positive dose-response. We have not used this method due to the many limitations of its interpretation. It is possible when grouping distant tissue sites to obscure an otherwise significant effect. This will occur if there are multiple tumors in several animals and some of the tumors have a high spontaneous probability of occurrence. In cases where there are marginally significant results in several tissues of different animals and there is sufficient information on the mechanism of action to justify grouping the tissues, it is possible to obtain a significant result. The information concerning the mechanism of action for 2,3,7,8-TCDD is insufficient to support this type of grouping.

To illustrate these points, consider the hypothetical example of a bioassay experiment with three dose groups and a control group and response at tissue 1 being 5/50 in all four groups. Now suppose tissue 2 is not examined as often as tissue 1 and the response in tissue 2 is 0/5 at control and 5/5 in the three dosed groups. If the 5 animals examined at tissue 2 also have a tumor at tissue 1, the combined response would be 5/5 in all groups and the rather marked effect in the second tissue would go undetected. This is only a sample of possible outcomes from grouping tissue sites. The point being stressed here is that tissue sites should be grouped only after careful review and that no automatic rule should be applied.

In applying these risk estimates to human populations, care must be taken in interpreting the variability of the results. Simulation studies (Portier and Hoel, 1983) and variance estimates have given us some indication of the variability of the statistical estimation process. However, with respect to the precision of the risk estimates for quantifying human response, there is no apparent theoretical method for measuring the biological errors that may occur. Several factors enter into the biological variability, such as absorption and metabolism, species' sensitivity, target tissue retention of the active compound, genetic variability, etc. Ideally, detailed information would be available on these factors, effectively reducing the error due to species differences.

In almost every case, however, this information is unavailable and the interspecies extrapolations are made in terms of species-equivalent dosage scales. The four most common conversion scales are (1) daily dose per kilogram body weight; (2) parts per million in the diet; (3) daily dose per body surface area; and (4) total accumulated dose. In converting effects from rats to humans, the total accumulated dose scale will yield the largest human risk estimates and the body weight conversion will yield the smallest human risk estimates. There is very little empirical evidence that would support the use of any one factor (NAS, 1975; Krasovskii, 1976; Rall, 1977; Hoel, 1979; Crouch and Wilson, 1979; Bernstein, Gold, Ames, Pike and Hoel, 1984).

There is some limited information on the mechanism through which dioxin effects the toxic response. First, it is unlikely that there is a significant amount of covalent binding of 2,3,7,8-TCDD to DNA (Poland and Glover, 1979). In addition, it is also unlikely that some metabolic product is involved. How to use this information in doing species scale-up is unclear. There is some evidence that 2,3,7,8-TCDD is slowly metabolized and doses are for the most part cumulative. This implies it may be prudent to convert from animal risk estimates to human risk estimates using a conversion based on the species specific target organ weights or target organ sizes and to use longer time units than one day.

CONCLUSIONS

In this paper we have attempted to portray a possible approach to the estimation of risks associated with exposure to 2,3,7,8-TCDD. 2,3,7,8-TCDD is carcinogenic in several species at several sites. After reviewing the available bioassay data, it was found that the female Sprague-Dawley rat was most sensitive. Using linear modelling, the estimated doses which would yield an added risk of one-in-one-million ranged from 38 fg/kg.bw/day in female rat livers to 3000 fg/kg.bw/day in female mice subcutaneous tissue. The statistical variability of these estimates has been addressed. However, it is not possible to determine if the assumption of linearity is valid. Also, it is not possible to quantify the error associated with the extrapolation of these results to humans. Even though the eventual mechanism of action is still unknown, the little that is known should be used in improving the species scale-up such as the absorption and distribution of 2,3,7,8-TCDD after exposure. In addition, studies on the mechanism of action of TCDD in the animals would greatly help to alleviate some of the uncertainty involved in species scale-up.

SUMMARY

Statistical methodologies for quantifying the risks from exposure to possible cancer-causing substances are applied to several animal bioassays on 2,3,7,8-tetrachlorodibenzo-p-dioxin (2,3,7,8-TCDD). The risks experienced by the study animals are estimated using both quantal response models and time-varying tumor prevalence models. In addition, a discussion of the advisability of grouping tumor sites is given.

REFERENCES

Albert, R. (Chairman). Risk assessment on (2,4,5-Trichlorophenoxy) Acetic Acid [2,4,5-T], (2,4,5-Trichlorophenoxy) Propionic Acid, and 2,3,7,8-Tetrachlorodibenzo-p-dioxin [TCDD]. Environmental Protection Agency, Washington, DC (September 12, 1980).

Armitage, P. and Doll, R. The age distribution of cancer and a multistage theory of carcinogenesis. British Journal of Cancer 8, 1-12 (1954).

Axelrod, D. Chlorinated hydrocarbons (U.S. Love Canal): Case studies of selected area wide environmental exposures, presented at the National Academy of Sciences Workshop on Plans for Clinical and Epidemiological Follow-Up after Area-Wide Chemical Contamination. Washington, DC (17 March, 1980).

Bernstein, L., Gold L., Ames, B., Pike, M. and Hoel, D. Some tautologous aspects of the comparison of carcinogenic potency in rats and mice. Submitted to Fundamental and Applied Toxicology, (1984).

Brown, K. and Hoel, D. Modelling time-to-tumor data: analysis of the ED01 study. Fundamental and Applied Toxicology 3, 458-469 (1983).

Chand, N. and Hoel, D. A comparison of models for determining safe levels of environmental agents. Reliability and Biometry, SIAM, Philadelphia, 681-700 (1974).

Crouch, E. and Wilson, R. Interspecies comparison of carcinogenic potency. Journal of Toxicology and Environmental Health 5, 1095-1118 (1979).

Crump, K., Hoel, D., Langley, C. and Peto, R. Fundamental carcinogenic processes and their implications for low dose risk assessment. Cancer Research 36, 2973-2979 (1976).

Crump, K. and Watson, W. GLOBAL79: A FORTRAN program to extrapolate dichotomous animal carcinogenicity to low dose. Science Research Systems, Ruston, LA (1979).

Hay, A. The Chemical Scythe: Lessons of 2,4,5-T and Dioxin. Plenum Press, New York (1982).

Hoel, D. Animal experimentation and its relevance to man. Environmental Health Perspectives 32, 25-30 (1979).

Hoel, D. Statistical measures of risk. Drug Metabolism Reviews 13(5), 829-838 (1982).

Hoel, D., Kaplan, N. and Anderson, M. Implication of nonlinear kinetics on risk estimation in carcinogenesis. Science 219, 1032-1037 (1983).

Kalbfleish, J. and Prentice, R. The Statistical Analysis of Failure Time Data. Wiley and Sons, New York (1980).

Kalbfleish, J., Krewski, D. and Van Ryzin, J. Dose response models for time to response toxicity data. Canadian Journal of Statistics 11, 25-50 (1983).

Kimbrough, R., Carter, C., Liddle, J., Cline, R. and Phillips, P. Epidemiology and pathology of a tetrachlorodibenzodioxin poisoning episode. Archives of Environmental Health 32, 77-86 (1977).

Kimbrough, R., Falk, H., Stehr, P. and Fries, G. Health implications of 2,3,7,8-tetrachlorodibenzodioxin (TCDD) contamination of residential soil. Journal of Toxicology and Environmental Health, 1984, in press.

Kociba, R., Keyes, D., Beyer, J., Carreon, R., Wade, E., Dittenber, D., Kalnins, R., Frauson, L., Park, D., Barnard, S., Hummel, R. and Humiston, C. Results of a two-year chronic toxicity and oncogenicity study of 2,3,7,8-tetrachlorodibenzo-p-dioxin (TCDD) in rats. Toxicology and Applied Pharmacology 46, 279-303 (1978).

Krasovskii, G. Extrapolation of experimental data from animals to man. Environmental Health Perspectives 13, 51-58 (1976).

Krewski, D., Crump, K., Farmer, J., Gaylor, D., Howe, R., Portier, C., Salsburg, D., Sielken, R. and Van Ryzin, J. A comparison of statistical methods for low dose extrapolation utilizing time-to-tumor data. Fundamental and Applied Toxicology 3, 140-160 (1983).

McConnell, E., Moore, J., Haseman, J. and Harris, M. The comparative toxicity of chlorinated dibenzo-p-dioxins in mice and guinea pigs. Toxicology and Applied Pharmacology 44, 335-356 (1978).

NAS. The effects of herbicides in South Vietnam, Part A, Summary and Conclusions. National Academy of Sciences, National Academy Press, Washington, DC (1974).

NAS. Pest control: An assessment of present and alternative technologies, Vol. 1, Contemporary pest control practices and prospects. National Academy of Sciences, National Research Council, Washington, DC (1975).

NAS. Risk Assessment in the Federal Government: Managing the Process. National Academy of Sciences, National Academy Press, Washington DC (1983).

NTP. Carcinogenesis bioassay of 2,3,7,8-tetrachlorodibenzo-p-dioxin in Osborne-Mendel rats and B6C3F1 mice. National Toxicology Program, Technical Report # 209, Research Triangle Park, NC (1982).

Poland, A. and Glover, E. An estimate of the maximum in vitro covalent binding of 2,3,7,8-tetrachlorodibenzo-p-dioxin to rat liver protein ribosomal RNA and DNA. Cancer Research 39, 3341-3344 (1979).

Portier, C. and Hoel, D. Low-dose rate extrapolation using the multistage model. Biometrics 39(4), 897-906 (1983).

Portier, C. and Hoel, D. Modelling time to response toxicity data: historical evidence for the adequacy of certain models. Unpublished manuscript (1984).

Rall, D. Species differences in carcinogenesis testing. In: Origins of Human Cancer, H. H. Hiatt, J.D. Watson and J.A. Winsten, Eds., pp. 1383-1390. Spring Harbor Laboratory, New York (1977).

Van Ryzin, J. and Rai, K. A dose-response model incorporating Michaelis-Menton kinetics. Columbia University Technical Report B-30, New York (1983).

HEALTH IMPLICATIONS OF 2,3,7,8-TETRACHLORODIBENZODIOXIN (TCDD) CONTAMINATION OF RESIDENTIAL SOIL

Renate D. Kimbrough, M.D.[1]
Henry Falk, M.D.[1]
Paul Stehr, Dr.P.H.[1]
George Fries, Ph.D.[2]

[1]Center for Environmental Health
U.S. Centers for Disease Control
Atlanta, Georgia 30333

[2]U.S. Department of Agriculture
Beltsville, Maryland 20705

This paper closely follows Dr. Kimbrough's presentation during the symposium. The complete report from which this paper is excerpted will be published by Hemisphere Publishing Corporation, Washington, DC 20005 in 1984. Adapted with permission of the publisher.

From Public Health Risks of the Dioxins, proceedings of a symposium held on October 19-20, 1983 at The Rockefeller University, New York City. Edited by William W. Lowrance. Published by William Kaufmann, Los Angeles, California, 1984.

INTRODUCTION

Much has been written about TCDD, and it is not the purpose of this document to review all available literature. For an overview, the reader is referred to recent reviews (Kimbrough 1980; National Research Council of Canada, 1981). Primarily, toxicology information useful for risk assessment will be reviewed here. In making risk assessments, investigators must consider the possible routes of human exposure and the average daily dose or the total lifetime dose. Although some adverse health effects have been observed in humans after exposure to TCDD, the dose these individuals received was not quantified. The only available dose-response data were obtained in animal studies. A critical feature of TCDD is that some animal species are much more susceptible to the toxic effects of TCDD than others. Since no dose-response information is available from human exposures, it is not known which animal species most nearly approximates human responses, although it appears that humans may not be the most sensitive species. From experience with other substances the guinea pig is often more sensitive than most other species.

Some scientists have even claimed that, with all of the exposure humans have received in occupational situations, humans must not be very susceptible to the toxic effects of TCDD, and the main health effect in humans is chloracne (May, 1973). Such far reaching statements seem inappropriate, however, since no information is available on how much TCDD the exposed workers who developed chloracne actually absorbed systemically and since chloracne can be either (1) part of a systemic disease or (2) produced locally by applying TCDD to the skin, with no toxic levels being absorbed into the body (Bauer et al., 1961). It is also too early to conclude what the long-term effects of chronic low-level exposure in humans will be. Furthermore, in estimating the risk that TCDD poses for the general population, we must consider children, females of childbearing age, the aged, and the infirm.

EXPOSURE ASSESSMENT AT TCDD-CONTAMINATED AREAS

Questions as to the habitability of any area where soil is contaminated with TCDD are necessarily linked to considerations of excess risks of developing specific adverse health effects as a result of the total cumulative dose an individual receives. In turn, this cumulative dose is a function of several factors:

1. concentrations of environmental contamination,
2. location of and access to contaminated areas,
3. type of activities in contaminated areas,
4. duration of exposure, and
5. specific exposure mechanisms.

Dose-rate may be an important factor in exposure assessment, but its effect on carcinogenic risk for dioxin is uncertain. The potential of increased risk from high doses at susceptible life stages may be balanced or exceeded by repair mechanisms operative at periods of lower doses. This cannot be ascertained based on currently available information. We have therefore based the exposure calculations in this section on cumulative dose averaged over a lifetime.

In addition, questions of continued habitability must also include considerations of the potential for limiting or eliminating ongoing exposures.

Developing a level of concern about an unacceptable risk due to exposure to TCDD poses significant difficulties because TCDD has such unique properties. (These are outlined in the complete version of this document.)

In the past, as a first approach, several groups have used a series of risk assessment estimates based on several of these factors to determine what an "acceptable" risk for exposure to TCDD would be. As more information on the toxicity of this chemical has become available, these levels have generally been reduced.

To determine whether a specific concentration of TCDD in soil presents a risk to humans, we must first examine how humans might absorb TCDD from such soil. Unfortunately, the amount of any chemical present in soil that may be absorbed by humans coming in contact with the soil is not well known. Most risk assessments that have been made in the past have been made for such media as food, where it is assumed that a certain amount of food with a certain concentration of the chemical in it is consumed; for air, where how much air is inhaled simply needs to be calculated; or for chemicals in water, where the only number needed is the amount of water consumed--although, as far as water quality criteria are concerned, the bioaccumulation of chemicals in fish from contaminated water is also considered. Unfortunately, the analogous series of estimates is more complicated for soil.

Basically, three exposure routes must be considered: dermal absorption through direct contact with the soil, ingestion of soil, and the inhalation of dust to which TCDD is attached. Vapors may be an additional, probably minor, route of exposure. Another issue, which does not directly enter into the current risk assessment, is the fact that TCDD in the environment could eventually end up in the food chain, particularly in fish. If TCDD enters a food chain, there is an unknown additional source of exposure which must be added to the risk of those individuals exposed to contaminated soil and of a larger, undefined population.

Regarding dermal absorption, there is some evidence that TCDD binds to the soil and would not be as easily available for absorption. (Vegetation covering contaminated soil may also decrease TCDD availability.) Information on bioavailability, however, is currently limited and may vary for different types of soil. According to the literature (Poiger and Schlatter, 1980, and a personal communication), anywhere from 1 to 10% of the TCDD in the soil may be absorbed through the skin, and this percentage is likely to depend on the TCDD concentration in the soil (i.e., it may be greater at higher concentrations) and on the type of soil. When Poiger and Schlatter (1980) applied soil with a dose of 350 ng TCDD to the backs of rats, 1.7 $\pm$ 0.5% of the dose was found in the liver; at a dose of 26 ng about 0.05% of the dose was found in the liver.

The authors do not state in the article how long the soil was left in contact with the skin of the rats except that after 24 hours it was not possible to recover all of the applied dose from the skin surface. Therefore, the subsequent estimates will allow for this range of skin absorption factors.

In regard to the portion of total dose due to ingestion of soil particles, feeding studies in animals suggest that 30% or more of the TCDD adsorbed on soil will be absorbed in the gastrointestinal tract (McConnell et al., 1984). Poiger and Schlatter (1980) found 16-24% of the administered dose of TCDD in the liver. According to Fries and Marrow (1975), this represents about 70% of the body burden of TCDD.

In regard to inhaled doses, little information is available on the amount of dust that may be present in the air in situations of known soil contamination; measurements in Seveso showed that the amount of dust in air was 0.14 mg/m^3 air (DiDomenico et al., 1980). No dust levels in air whose sole source is soil are available from air monitoring stations. Soil, vegetable matter, and particles from other sources such as car exhaust are measured as particulate matter. The use of particulate matter would highly overestimate dust derived from soil. In riding arenas or in relatively drier areas, dust levels would be possibly higher. On the other hand, immediately after a rainfall there would probably be less dust. In the same investigation, it was shown that TCDD levels in dust were comparable to those found in soil. Another unknown is the amount of material that could be carried into the house from the outside. It is conservatively assumed that the exposure to dust inside a house surrounded by contaminated soil is similar to that which would occur if people spent their entire time in contact with the contaminated soil outside. (One of the CDC consultants commented that the assumption that indoor levels will equal outdoor levels appears unnecessarily conservative.) An average adult at rest exchanges approximately 10 m^3 of air per 24-hour period and this increases with mild activity to 18-24 m^3/day and to 40 m^3/day with hard physical labor. Finally, it is assumed that whatever TCDD is inhaled adsorbed to dust particles is absorbed either through deposition in the respiratory tract or by ingestion after being brought up by the ciliary action of the respiratory tract epithelial cells.

Several comments were received from CDC consultants on exposure estimates. (e.g., How much soil does a young child eat when playing outside? How much soil gets onto the skin during gardening activities? How much soil gets onto the skin of children playing football or other games?) Unfortunately, there is no documentation in the literature which clarifies the problems raised. To illustrate, 1 g of soil less than 1 mm thick can be spread over an area of 4-5 cm^2 (or 1.5-2 $inches^2$). Ten g of soil less than 1 mm thick can be spread over an area of about 15 cm^2 (or about 6 $inches^2$). The volume of dirt will vary somewhat with moisture content. The soil used in calculating the above surface areas was Georgia clay that had been stored for several months at room temperature. Thus the surface area of one medium-sized adult's or child's palm was computed to come in contact with 1-10 g of soil.

All of the calculations regarding exposure are based on the assumption that humans have access and contact with the contaminated soil and that a percentage of the TCDD present in the soil is absorbed. The frequency of access and contact must also be considered, and for dermal exposure it must be remembered that clothing will afford some protection. The doses calculated below are, in some sense, worst-case estimates for the concentrations used.

A large number of estimated total daily doses can be derived from the many combinations of the exposure route-specific doses (given different sets of assumptions as to absorption rates, soil contamination, etc.). However, in an attempt to derive a more accurate estimate of exposure to and uptake from contaminated soils, a simulation analysis was performed utilizing a further set of assumptions which are most likely obtained in reality. For instance, TCDD is suggested to have an environmental half-life of approximately 10-12 years. Therefore, the simulation model assumed that soil contamination levels would decrease in a log-linear fashion with a 12 year half-life following the equation

$$TCDD_t = e^{(a - 0.00016 \times t)}$$

where: $TCDD_t$ = current soil level on day "t" (in ppb)
a = ln (initial soil contamination level in ppb)
t = number of days elapsed since initial soil contamination measurement.

Consideration of the route-specific uptakes does not require an assumption of constant exposure but merely postulates an *a priori* reasoned absolute daily amount of soil contact. For inhalation, we assumed that airborne dust is contaminated at the same level as surrounding outside soil and that 15 m^3 air is exchanged per day. In regard to ingestion, based on preliminary results of bioavailability studies, a GI absorption rate of 30% from soil seems most tenable. In addition, the amount of soil which the average person is likely to ingest will be dependent on characteristic activity patterns which, in turn, are closely dependent on age. Based on work done studying lead uptake from contaminated soils, a reasonable pattern of soil ingestion (which still maintains conservative estimates) for specific age groups was constructed and is presented in Table 1.

Table 1. Daily soil ingestion patterns by age.

Age Group	Soil Ingested
0-9 months	0 grams
9-18 months	1 gram
1.5-3.5 years	10 grams
3.5-5 years	1 gram
>5 years	100 milligrams

Based on animal experimentation, the best estimate for dermal absorption of TCDD is approximately 1% (especially in the low ppb contamination ranges). Similar reasoning for an age-dependent pattern of deposition of soil on exposed skin as for soil ingestion was used to derive Table 2.

Table 2. Daily deposition of soil on skin by age.

Age Group	Amount on Skin
0-9 months	0 grams
9-18 months	1 gram
1.5-3.5 years	10 grams
3.5-15 years	1 gram
>15 years	100 milligrams

Finally, it was assumed that these exposures would probably take place only six months of the year because of seasonal influences and varying activity patterns.

The simulation model was run using a computerized iterative procedure by daily increments to estimate total lifetime dose using the following formula:

$$\text{Total Lifetime Dose} = \sum_{t=1}^{T} TCDD_t \times (ING_t \times GI + DERM_t \times ABS + INH \times DUST) \times SEAS$$

where:
- T = expected lifespan (in days)
- ING_t = age-specific amount of soil ingested at time "t"
- GI = % absorbed through gastrointestinal tract
- $DERM_t$ = age-specific amount of soil deposited on skin at time "t"
- ABS = % absorbed through skin
- INH = amount of air exchanged per day
- $DUST$ = concentration of dust in air
- $SEAS$ = "dummy" variable for seasonal access to outdoor contaminated areas (i.e., = 1 for fair weather months and = 0 for cold weather months).

For the sake of brevity, extreme total daily dose estimates were compiled for two divergent levels of TCDD soil contamination (1 ppb and 100 ppb). Based on the assumptions set forth above, at an initial soil contamination of 1 ppb in residential areas, the average daily TCDD dose to an individual over a 70 year lifetime would be 44.6 picograms. This is equivalent to 636.5 fg/kg b.w./day for a 70 kg person. For soils initially contaminated at 100 ppb, the average daily dose is estimated at 4.5 nanograms or 63.7 pg/kg b.w. for a 70 kg individual (Figure 1).

Figure 1. Estimated average daily dose corresponding to initial TCDD soil contamination levels.

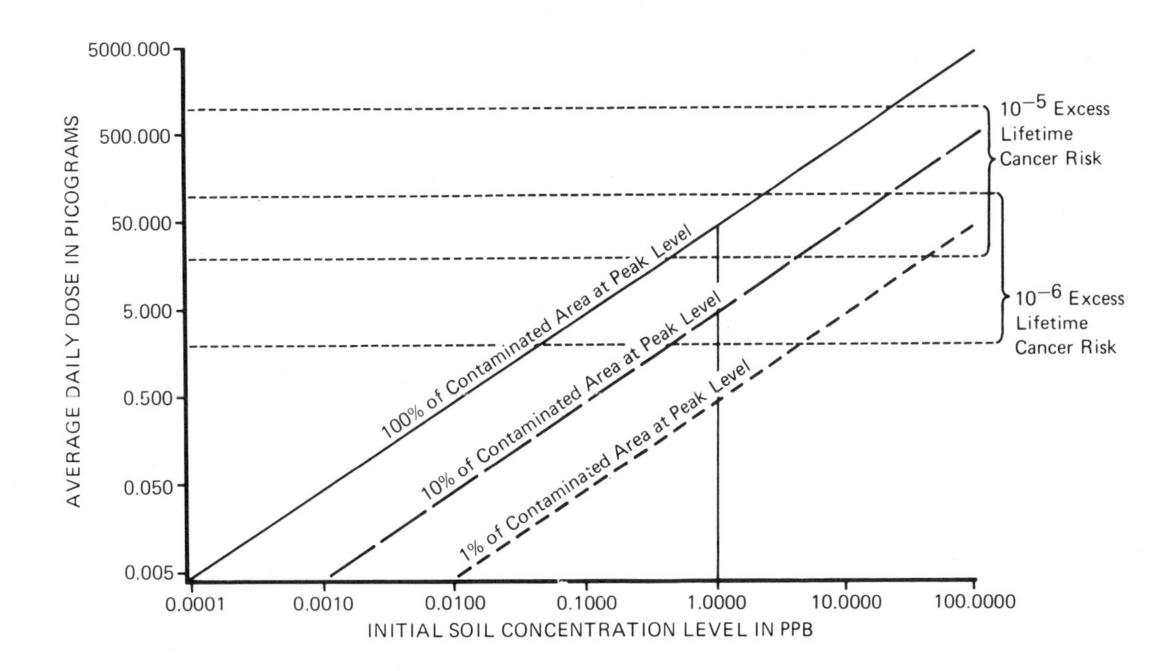

This figure illustrates the average daily dose of TCDD which would be received if 100, 10, or 1% of the accessible soil were contaminated with the peak recorded level of TCDD. The boundaries for an excess lifetime cancer risk for 10^{-5} and 10^{-6} are also illustrated.

In addition, it must be recognized that not all soil is contaminated uniformly in a given area. At the 1 ppb level the dose (and the resulting risk) would be considerably reduced if only 1% or 10% of the accessible soil contained TCDD at concentrations of 1 ppb and the rest of the soil were not contaminated (illustrated in Figure 1).

RISK ASSESSMENT

The critical step in assessing individual risks at these estimated dose levels must incorporate a comparison to known (or estimated) "safe" levels of exposure in relation to clearly defined health effects endpoints.

The National Research Council of Canada (NRCC) has recently published a report reviewing available toxicity data for TCDD and related compounds and various procedures for calculating a "virtually safe dose" (VSD) for TCDD from such data. A summary table of this document listed the various models, estimated VSDs, approximate 95% confidence levels, and references to the different models which were employed. This table has two typographical errors. The virtually safe dose differs from other calculations because only liver tumor data and body weight instead of body surface were used. The analyses in this section follow guidelines recommended by the CDC consultants and are based upon two studies: one by Kociba et al. (1978) and another by the National Toxicology Program (NTP, 1982). Further details on these (and other) studies can be obtained from the EPA (U.S. EPA, 1980). All of the CDC consultants agreed that the available human data are inadequate to be used in risk assessment calculations.

In the Kociba study a substantial proportion of the animals, including those in the control group, died before the 2-year sacrifice (78% to 92% in the males and 68% to 92% in the females). In addition, there appeared to be time-related and dose-related effects at the lesion sites. Results from a time-adjusted analysis of these data did not differ markedly from results reported in this document (C. Portier et al., this symposium).

The important lesion sites (for risk assessment) in the two sexes are given in Table 3, along with the tumor incidence at each dose.

Table 3. Tumor incidence at important lesion sites in the Kociba TCDD study.

Site	Sex	Pathologist	Control	Tumor Incidence at Administered Dose 0.001	0.01	0.1
Stratified squamous cell carcinoma of the tongue	Males	Kociba	0/76	1/49	1/49	3/42
		Squire	0/77	1/44	1/49	3/44
Nasal turbinates/ hard palate squamous cell carcinoma	Males	Kociba	0/51	0/34	0/27	4/30
		Squire	0/55	1/34	0/26	6/30
	Females	Kociba	0/54	0/30	1/27	4/24
		Squire	0/54	0/30	1/27	5/22
Hepatocellular nodules and carcinoma	Females	Kociba	9/86	3/50	18/50	34/48
		Squire	16/86	8/50	27/50	33/47
Lung keratinizing squamous cell carcinoma	Females	Kociba	0/86	0/50	0/49	7/49
		Squire	0/86	0/50	0/49	8/47

The original pathology done by Kociba was reviewed by Squire (U.S. EPA, 1980), and the results of Squire's review are therefore also included. In the EPA's risk assessment, the agency based its analysis upon grouping these sites, using as incidence the number of animals with any one of the lesions divided by the number of animals examined at any of these sites for each sex. We have not used this procedure; instead, the multistage model (Crump et al., 1977) was fitted to the tumor incidence from each lesion site. These results appear in Table 4, along with the chi-squared value for the goodness-of-fit test.

Table 4. Estimates and approximate 95% lower confidence limits for the VSD[1] of TCDD from the Kociba study in Sprague-Dawley rats.

Lesion/Pathologist	Model[2]	Added Risk 1.E-4		Added Risk 1.E-6		Chi-squared G-O-F[4]
Male rat stratified squamous cell carcinoma of the tongue						
Kociba	Linear*	142581	(59480)	1426	(595)	1.43
Squire	Linear*	151793	(62642)	1518	(626)	1.60
Male rat nasal turbinates or hard palate squamous cell carcinoma						
Kociba	Cubic*	8877014	(17493)	1912466	(175)	0.00
	Linear	77507	(37566)	775	(376)	0.39
Squire	Cubic*	7757813	(11676)	1671343	(117)	2.39
	Linear	49468	(26431)	495	(264)	2.72
Female rat lung keratinizing squamous cell carcinoma						
Kociba	Cubic*	8659951	(32477)	1865707	(325)	0.01
	Linear	72600	(41318)	726	(413)	0.82
Squire	Cubic*	8126063	(26834)	1750685	(268)	0.01
	Linear	60375	(35490)	604	(355)	0.98
Female rat nasal turbinates or hard palate squamous cell carcinoma						
Kociba	Linear*	49771	(25812)	498	(258)	0.49
Squire	Linear*	37237	(20313)	372	(203)	0.20

Table 4. Continued.

Lesion/Pathologist	Model[2]	Added Risk 1.E-4		1.E-6		Chi-Squared G-O-F[4]
Female rat liver hepatocellular carcinoma or adenoma						
Kociba	Linear*	7742	(5725)	77	(57.2)	6.35
	Transf[3]	3836	(2919)	38	(29.1)	2.18
Squire	Linear*	8649	(6074)	86	(60.7)	10.36
	Transf[3]	3760	(2763)	38	(27.6)	4.30

[1]Entries are virtually safe dose (VSD) Linear-cubic lower confidence bounds (LCB) in fg/kg b.w./day by using the multistage model.

[2]Linear Model: P(d)=1-e(-a-b*d) Cubic Model: P(d)=1-e(-a-c*d**3).

[3]A multistage model is fitted to the average liver concentration of TCDD, and safe doses are transformed by using the relationship:

Administered dose = Liver dose/510.297

The divisor is the least-squares slope between the administered and the average liver dose of TCDD for control and the two lowest doses.

[4]G-O-F = goodness of fit.

*best fit.

When the best fit to the data was nonlinear, the linear model was also fitted (see the note at the bottom of Table 4 for a description of this model) and produced risk estimates for comparison. In all sites, except female liver tumors, the data could be adequately fitted. For males, the smallest lower confidence bound on the "virtually safe dose" (VSD) for an added risk of 1/1,000,000 is 117 fg/kg b.w./day. In females, the most sensitive site seemed to be the liver, but it was not possible to get an adequate fit using the administered dose. In the original manuscript, Kociba et al. (1978) had determined the concentration of TCDD in the livers of a sample of the sacrificed animals from each dose group. The means for each dose group appear below:

Administered dose	0.001	0.01	0.1
Liver concentration (ppb)	0.540	5.10	24.0

Assuming these concentrations were present in the animals at stable levels for much of the study, these are the appropriate doses to which the liver tumor incidence data should be fitted. Assuming the relationship between administered dose and liver dose is nonlinear above the 0.01 ug/kg b.w./day dose and linear below this dose (as appears to be the case), liver dose can be transformed back to administered dose by using the least-squares line through the points (0,0), (0.001,0.54), and (0.01,5.1). This leads to the linear relationship

Administered dose = Liver dose/510.297.

However, the data on liver concentrations may not be sufficiently precise to assume dose dependent linear or non-linear relationships. The VSD estimates and lower confidence bounds in the administered-dose scale appear in Table 4 under model type "Transf." With this approach, the smallest lower confidence bound on the VSD by using an added risk of 1/1,000,000 cancers is 27.6 fg/kg b.w./day for female rats.

The NCI/NTP study (1982) was a gavage experiment on B6C3F1 mice and Osborne-Mendel rats, of both sexes. Seventy-five vehicle-treated control animals and 50 animals were treated at each of three doses for each sex by species combination. The doses were administered twice weekly. Weekly doses were divided by seven; thus daily doses were obtained and the same scale could be used. There are, of course, questions of peaks and dips in body content of TCDD. We have assumed that on a weekly scale the dose is approximately constant and that division by 7 to yield daily doses is an acceptable conversion. Both sexes in rats and male mice received doses of 0.0014, 0.0071, and 0.0714 ug/kg b.w./day (0.01, 0.05, and 0.50 on the weekly dose scale). Female mice received doses of 0.0057, 0.0286, and 0.2859 ug/kg b.w./day (0.04, 0.2, and 2.0 on the weekly scale). There were no significant survival differences in any group and, in fact, the estimates of the VSD based upon a time-to-tumor model (multistage Weibull) were similar to the estimates obtained from the linear model (Crump et al., 1977).

The important lesion sites, the estimates of risk, and the chi-squared goodness-of-fit statistic are given in Table 5.

Table 5. Estimates and approximate 95% lower confidence limits for the VSD[1] of TCDD from the NCI/NTP study.

Lesions	Model[2]	Added Risk 1.E-4	Added Risk 1.E-6	Chi-Squared G-O-F[3]
Male rat thyroid				
Follicular-cell adenoma	Linear*	40291 (21435)	403 (214)	4.81
Female rat thyroid				
Follicular-cell adenoma	Cubic*	7142868 (36711)	1542868 (367)	0.48
	Linear	75737 (34895)	757 (349)	0.79
Female rat liver				
Neoplastic nodules	Lin-Quadratic*	1442863 (19298)	105571 (193)	1.31
	Linear	31520 (18784)	315 (188)	1.61
Nodules and carcinomas	Quadratic*	1338200 (16523)	133820 (165)	1.31
	Linear	25552 (16001)	256 (160)	1.74
Male mouse liver				
Hepatocellular carc.	Linear*	26293 (15147)	263 (151)	1.01
Adenomas and carc.	Linear*	13394 (8633)	134 (86)	0.14
Female mouse liver				
Hepatocellular carc.	Linear*	246021 (121594)	2460 (1216)	0.74
Adenomas and carc.	Linear*	145013 (77876)	1450 (779)	2.59
Female mouse thyroid				
Follicular-cell adenoma	Linear*	301051 (139090)	3011 (1391)	3.75

Table 5. Continued.

Lesions	Model[2]	Added Risk 1.E-4	Added Risk 1.E-6	Chi-Squared G-O-F[3]
Female mouse sarcoma subcutaneous tissue	Lin-Cubic*	430263 (142803)	4303 (1428)	0.05
	Linear	295289 (142777)	2953 (1427)	0.06
Female mouse lymphoma and leukemia	Linear*	102852 (54271)	1028 (543)	0.02

[1]Entries are virtually safe dose (VSD) lower confidence bounds (LCB) in fg/kg b.w./day by using the multistage model.

[2]Linear: P(d)=1-3(-a-b*d) Quadratic: P(d)=1-3(-a-c*d**2)

Cubic: P(d)=1-3(-a-f*d**3) Lin-Quadratic: P(d)=1-3(-a-b*d-c*d**2)

Lin-Cubic: P(d)=1-3(-a-b*d-f*d**3)

[3]G-O-F = goodness of fit.

*best fit.

As before, when the linear model was not the best fit, it was fitted separately in order to see what difference this model would make. All of the models gave acceptable fits to the data. Where there were two or more lesion sites for a particular animal group, the EPA pooled the results as mentioned before. Again, independent sites were not combined.

The smallest lower confidence bounds on the VSD (1.E-6) for each sex/species combination are as follows: male rat thyroid, 214; female rat liver nodules and carcinomas, 160; male mouse liver adenomas and carcinomas, 86; and female mouse lymphoma/leukemia, 543. These results do not differ markedly from the results of the Kociba study.

Thus, the risk assessment calculations for the different tumors in these two studies provide a dose range of 280 fg/kg b.w./day to 14 pg/kg b.w./day that would result in an increased cancer risk of 1/100,000, and a dose range of 28 fg/kg b.w./day to 1428 fg/kg b.w./day that would result in an increased cancer risk of

1/1,000,000. Direct conversions from rodents to humans were used. It is not clear whether this is justified. Humans repair DNA faster than rodents, and many other differences could be pointed out. None of the extrapolation approaches has at present absolute indisputable scientific justification, including the approach used in this document. However, there is also no scientifically justified alternative form of extrapolation (e.g., use of safety factors) that should preferentially be used. Furthermore, body weight instead of body surface was used, since sizes of organs and, thus, doses to organs correlate better with body weight. Many other biological factors also correlate better with body weight (Krasovskii, 1976).

These calculations assume that a linear dose-response relationship exists for carcinogens that are believed to be primarily promoters. The dose-response curve for promoters may, however, not be linear, causing an overestimate of the risk. The model was used on a hypothetical basis and the cancer risk for TCDD should be reevaluated as the data base enlarges. The linear dose response relationship, however, has not been shown experimentally, and it is not understood how promoters affect cancer growth at very low concentrations. TCDD apparently affects cell membranes through lipid peroxidation (Stohs et al., 1983), which also affects membrane fluidity. The effect of TCDD on membranes is a likely reason for the formation of multinucleated cells in TCDD-exposed animals (Jones and Butler, 1974); Kimbrough et al., 1977). Most likely, at very low dosage levels, antioxidants such as vitamin E, vitamin C, selenium, and unsaturated fatty acids would have a protective effect against the promoting actions of TCDD (Ames, 1983). Certainly many other naturally occurring and man-made materials also cause lipid peroxidation (Ames, 1983). At doses of TCDD in the picogram range, the contribution that this chemical would make might be overridden by the effect of this potpourri of substances and modified by dietary intake of natural antioxidants and thus may not greatly contribute to the risk of cancer or other chronic diseases.

Sweeney et al. (1979) have shown that iron deficiency results in decreased liver toxicity of TCDD. It has also been shown with related chemicals, such as the polybrominated biphenyls, that serosal transfer of iron from duodenal gut sacs of rats is increased. This also occurs in vivo in rats with a concomitant rise in serum iron (Manis and Kim, 1980). It is possible that TCDD would have similar effects on iron absorption. For these reasons and, as illustrated by the receptor model developed by Poland, it is not known whether the linear-derived multistage model for assessing cancer risk is the most appropriate. Unfortunately, a scientific data base which would permit the use of different, possibly less conservative, models does not exist. Although a different model could be developed based on the receptor theory, it is not entirely clear that all of the toxicity of TCDD is controlled by the Ah locus (Sweeney and Jones, 1983).

Since the no-observable-effect levels for reproduction, immune toxicity, and various other toxic effects are not established in various species, a conservative approach for chronic toxicity in general is in order. The study by Murray et al. (1979) suggested that 0.001 ug/kg/day is a no-observed effect level for reproduction in rats. This study shows a very varied fertility index among the controls through different generations; in addition, TCDD body burdens of the dams are greatly affected by lactation, introducing another variable. CDC consultants Drs. Hoel, Van Ryzin, and Portier also reviewed these data and concluded that there was insufficient evidence for an effect at 0.001 ug/kg b.w./day. For these reasons, this study was not used for risk assessment calculations (see also Appendix II in the complete review), and only the chronic toxicity studies which demonstrated a carcinogenic response in rodents were used.

Subhuman primates (which are much more susceptible to the effects of TCDD) show an effect on reproduction if fed for 6 months at a daily dose of 1.8 ng/kg. If the toxicology data from subhuman primates are used, then a 1000-fold safety factor would have to be used, since the lowest dose of 1.8 ng/kg per day was not a no-observed-effect level and was not obtained from a chronic feeding study. Thus, a daily dose rate of 0.0018 ng/kg--corresponding to a total daily dose of 144 pg--would be tolerable for an 80-kg person. For a 60-kg person, the tolerable total daily dose would be 108 pg. Thus, at the lowest daily dose likely to be obtained as estimated above for a soil level of 1 ppb (111.4 pg/day), both of these extrapolations from reproductive studies in animals appear to suggest a situation of no excess risk in humans. However, at virtually all other estimated levels of daily dose (i.e., under more severe sets of assumptions or the higher level of TCDD in soil), one might expect adverse reproductive health effects.

Figure 2 illustrates the lower and upper bounds of the virtually safe dose for a lifetime excess cancer risk.

Figure 2. Excess lifetime risk of developing cancer expected from various initial TCDD-soil contamination levels.

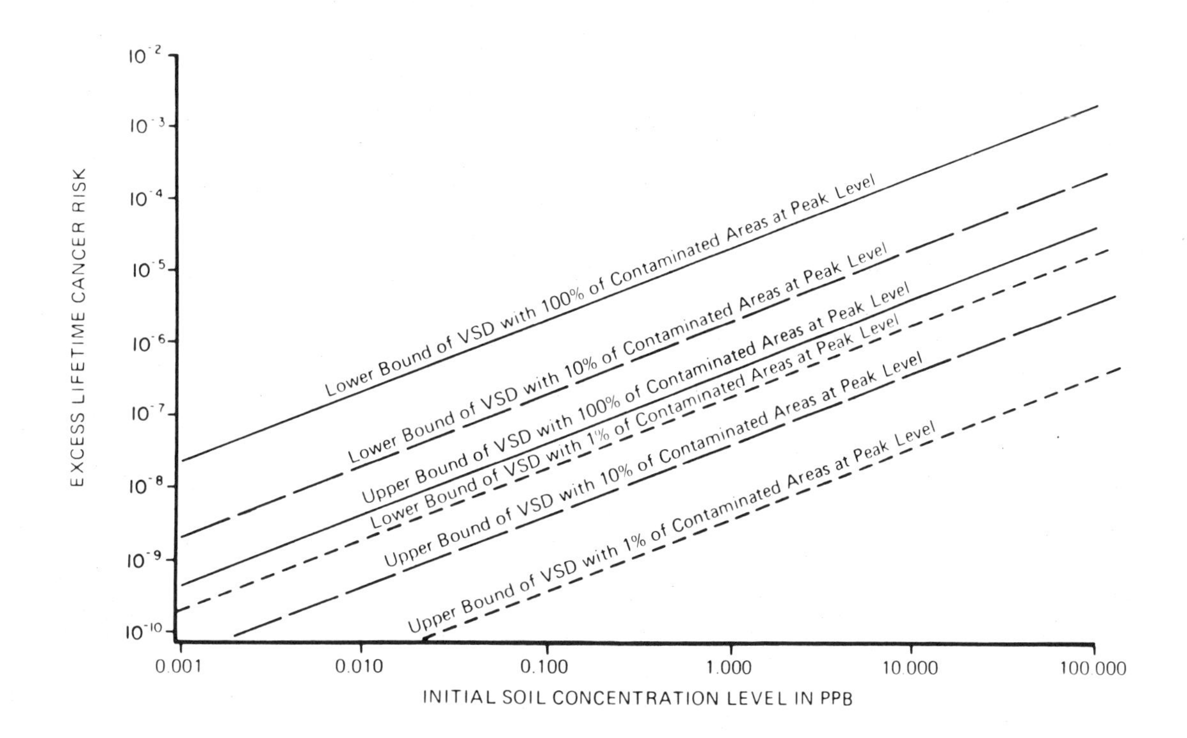

This information was derived from the calculations developed from animal data (Tables 4 and 5). At the concentration of 1 ppb, if 100% of the accessible soil contained TCDD at this concentration, the area of risk which is bordered by the upper and lower bound of the virtually safe dose does not represent an unacceptable cancer risk given the fact that the background cancer incidence in the general population is of a much higher order of magnitude. If less than 100% of the soil was contaminated this risk would even be further reduced. However, at levels much above 1 ppb the risk would become unacceptable.

As shown in Figure 2, the estimate for an increased cancer risk of 1 x 10^{-6} over a 70-year lifetime lies in a range bound by an upper dose (1.4 pg/kg b.w. per day) and a lower dose (28 fg/kg b.w. per day). Even if 100% of the soil were contaminated with 1 ppb TCDD, the highest daily dose an individual theoretically could receive would be 637 fg/kg b.w., which would be below the 1.4 pg/kg dose but above the 28 fg daily dose. For the estimates illustrated in Figure 2, the excess lifetime cancer risk from exposure to residential soil with a peak TCDD contamination level of 1 ppb ranges over 4 orders of magnitude, from above 10^{-5} to below 10^{-8}. At the lower bound for the VSD of 28 fg/kg b.w./day (or 1.960 pg/70-kg person/day), it would take just over three years to accumulate a total dose sufficient to increase an individual's lifetime risk of developing cancer by one in a million (using the estimates derived herein and assuming 100% of the contaminated area is at the peak recorded level). Over a 70 year lifetime, this would amount to a 0.000023 absolute increase (equivalent to 0.01% relative increase) over one's "normal" 25% probability of developing cancer in the United States (RR = 25.0023/25 = 1.00001).

It must be stressed that the exposure assessments used in estimating risks for carcinogenicity and reproductive health effects are based on critical assumptions that are not likely to be actually encountered. Most prominent of these is the assumption of uniform levels of contamination throughout the living space. In fact, areas with elevated TCDD levels are likely to be found in specific, well-defined locations that have concomitant unique use and access characteristics. Therefore, in an area where access is less than total, the actual daily exposure will be lower. Similarly, different usage patterns of affected areas (e.g., sports activities, gardening, horseback riding) or an individual's characteristics are not likely to lead consistently to worst-case situations and will have differing effects on the determination of total cumulative dose. It could be further argued that the daily dose rate is more important than the cumulative total dose. It appears, however, that exposure has to be for a sufficiently long time for cancer to develop since short exposure periods may result in recovery (Farber, 1974), or would not appreciably increase the cancer risk over background (Office on Smoking and Health, 1982). We have therefore concluded that a soil level of 1 ppb TCDD in residential areas is a reasonable level at which to express concern about health risks.

IMPLICATIONS FOR RISK MANAGEMENT

Therefore, where residential soil exceeds 1 ppb TCDD, risk management decisions on habitability and limiting exposure may range from recommendations to avoid identified "hot spots"; to limit specific activities in these areas (if possible); to temporarily relocate while contamination clean-up and onsite stabilization operations are performed; to permanently relocate and impose access restriction for a given site. In addition, such recommendations will have to be prepared in terms of situations which range from the need for near-term action to those of a less urgent nature. In all of these scenarios, however, these decisions must be made on a site-specific basis, as indicated by the complexities and variability of circumstances.

Although levels of TCDD below 1 ppb are, for practical purposes, not considered levels of concern, several additional considerations related to the risk assessment calculations should be pointed out to decision makers involved in risk management:

1. The calculations for a 1 ppb level and the judgment that this is a reasonable level for concern assume that exposure probably will not be consistent for a lifetime, since TCDD will slowly degrade, and that people will not be exposed extensively to the soil

on a continuous daily basis. For instance, it is anticipated that during cold or rainy weather, not much outside activity will occur. From the limited available information, it also appears that levels of TCDD within houses are at least 100-times less than levels measured outside the houses.

2. The precise bioavailability of TCDD from soil is not known. Such bioavailability may vary with the soil type. It has recently been established that TCDD contaminated soil from Missouri is toxic to guinea pigs and rats, if given orally. It was estimated that the bioavailability was 30-50% or more (McConnell et al., personal written communication, 1983).

3. The recovery of TCDD that is extracted from soil for chemical analysis varies a great deal and may be as low as 20%.

4. Fries and Marrow (1982) have shown that cattle, sheep, and swine consume up to 7% soil/day in their total ingested dry matter when grazing on ranges. Judging from experience with polybrominated biphenyls and 1,2,3,6,7,8-hexachlorodibenzodioxin, levels in the adipose tissue of these animals will probably bioaccumulate (see risk assessment scenario for "Ranges"). Soil levels of TCDD on ranges and other farmland should, therefore, reflect the levels given in Table 6.

Table 6. Concentrations of TCDD in soil that are projected to produce the maximum allowable residues in foods.

Food	TCDD in fat, pg/g	Observed Ratio[1]	Soil, pg/g
Beef[2]	7.9	0.39	20
Beef(cull dairy)[3]	7.9	0.10	79
Pork	22.7	1.86	12
Milk	2.5	0.40	6.2

[1]Concentration of polybrominated biphenyls (PBB) in product/ concentration of PBB in soil (Fries and Jacobs, 1983).

[2]Includes dairy cattle that have never lactated.

[3]Older cows. Younger cows would approach the values for beef cattle.

5. Furthermore, if contaminated soil is close to waterways and can contaminate these waterways by way of erosion, acceptable levels may also have to be lower, since fish can bioconcentrate TCDD 20,000-fold or more (National Research Council of Canada (NRCC), 1981). The Food and Drug Administration (FDA) has set concern levels for fish; at 50 ng TCDD/kg edible portion, fish should not be consumed, and at 25-50 ng/kg, fish should not be consumed daily. [Note: The FDA concern levels were established prior to work on this document and reflect both theoretical and practical considerations.]

SCENARIOS

Environmental situations may vary widely, and whether a certain level of TCDD in soil will give rise to concern has to be evaluated on a case-by-case basis. To provide guidance for those who have to deal with these situations, we describe several scenarios of how a level of concern is developed that is appropriate to the specific situation.

The calculations made in Tables 1 and 2 are for residential areas and include children.

Exposure in factories would usually be for a 40-hour work week. This situation will be addressed separately by the National Institute for Occupational Safety and Health.

Other areas considered are commercial areas, farmland, and areas that are essentially remote or inaccessible. Conditions in individual situations may vary considerably and in the end require judgment and common sense.

Commercial areas. At most commercial sites, parking lots and sidewalks are usually either paved or graveled, and exposure of the general population is transient and does not occur every day. Usually, children do not play in such areas for any length of time. In such situations, ingestion would be negligible, and inhalation and dermal exposure would be the primary routes. When concrete or gravel is involved, inhalation would also be negligible; thus, dermal absorption would be the primary exposure route. Since these total doses for all routes are so much smaller than in residential areas, a level of concern may not necessarily be reached unless levels are several fold or above 1 ppb. However, if concentrations are high or if there is a possibility for movement of soil or a dust problem, remedial measures (such as paving of the area) should be considered in any event. If levels are very high, e.g., above 100 ppb, more extensive remedial action may have to be considered.

Farmland. As stated above, the uptake of TCDD from soil by plants is negligible for most plants that have been tested. (This statement does not refer to instances of direct application of TCDD to plants or crops.) However, erosion into streams may occur. This might be prevented by leaving several feet at the edge of the land unplowed and planting grass, shrubs, and/or trees on this edge.

Ranges. The situation is different for ranges where cattle graze, as outlined below. [Note: This section is based on information and guidance from the U.S. Department of Agriculture, which is primarily involved in the evaluation of exposures in this setting.]

In this brief assessment, guidance is provided on concentrations of TCDD in soil that would be of concern when animals exposed to that soil are used to produce foods for human consumption. Farm families that use milk from their own cows or slaughter their own animals for home consumption would probably receive the highest exposures to TCDD from food (excluding fish). If the products were consumed away from the farm, exposures to the general population would be much lower because they would be diluted with uncontaminated products. This assessment is made for a maximum allowable intake of 100 pg TCDD/day. Proportional adjustments can be made in the assessment for changes in allowable intake or other assumptions because there is no evidence that the transfer processes for TCDD are affected by concentration.

The assumptions used for average daily beef, pork, and milk consumption are in Table 7. The values for beef and pork are derived from values for U.S. *per capita* retail sales of these meats in the past 5 years. *Per capita* consumption of veal and lamb is too low to merit serious consideration here. The value used for milk is higher than *per capita* consumption, but it is a value that has been used frequently in risk assessment, and it is reasonable for some segments of the population such as teenage boys. The fat contents, daily fat intakes, and allowable concentrations of TCDD to maintain intake under 100 pg/day for each of these foods are also in Table 7.

Table 7. Daily consumption, fat content and allowable concentration of TCDD in fat to maintain human TCDD intake below 100 pg/day/person.

	Consumption, g/day	Fat,%	Fat consumption, g/day	Allowable TCDD in fat, pg/g[1]
Beef	105	8-12	8.4-12.6	7.9
Pork	54	6-8	3.2-4.3	22.7
Milk	1000	4	40	2.5

[1]Based on the higher fat intake value.

Data directly relating soil concentrations of TCDD to concentrations of TCDD in milk or tissues of exposed animals are lacking, and these values must be estimated by indirect means. Firestone et al. (1979) fed pentachlorophenol (PCP) that contained several dioxins to lactating cows for 70 days. The PCP did not contain TCDD, but milk and tissue residue values were obtained for several other dioxins. The most efficiently absorbed dioxin was 1,2,3,6,7,8-hexachlorodioxin, and its concentration in milk fat was 2.4 times the concentration in the diet at the end of the study. This ratio is much lower than the ratios for many chlorinated hydrocarbons (Fries, 1982), and it is somewhat lower than the 3.1-to-1 ratio for polybrominated biphenyls (PBB) under similar experimental conditions (Fries and Marrow, 1975). Therefore, the results of a study of animal residues on Michigan farms with residual PBB in soils (Fries and Jacobs, 1983) can be used as a reasonable model for assessing movement of TCDD from soil to animals.

The ratios of concentrations of PBB in milk and tissue fat to concentrations in soil are in Table 6. When the ratios are divided into the maximum allowable TCDD in fat, soil concentrations are obtained that might be expected to produce the maximum allowable concentration in food products. The most serious problem in this assessment is with swine, which consume more soil than other species. Among ruminants, the most serious concern would be from animals that have never lactated.

The values in Table 6 were derived for situations in which animals were confined to limited holding areas. Direct measurements of the type used here are not available for animals on pasture. However, a review (Fries, 1982) of soil ingestion by animals on pasture suggests that the situation would not differ markedly under most conditions, but that milk would become relatively more important as a source of residues.

The values derived in this assessment should not be considered absolute determinants of the suitability of soils for livestock production. Rather, the value should be considered a guide to situations that require a more detailed evaluation.

REFERENCES

Ames, B.N. 1983. Dietary carcinogens and anticarcinogens: Oxygen radicals and degenerative diseases. Science 221:1256-1264.

Bauer, H., Schulz, K.H., and Spiegelberg, U. 1961. Berufliche Vergiftungen bei der Herstellung von Chlorphenol-Verbindungen. Archiv fur Gewerbepathologie und Gewerbehygiene 18:538-555.

Crump, K.S., Guess, H.A., and Deal, L.L. 1977. Confidence intervals and test of hypothesis concerning dose response relations inferred from animal carcinogenicity data. Biometrics 33:437-51.

DiDomenico, A., Silano, V., Viviano, G., and Zapponi, G. 1980. Accidental release of 2,3,7,8-tetrachlorodibenzo-p-dioxin (TCDD) at Seveso, Italy. Ecotoxicology and Environmental Safety 4:346-356.

Farber, E. 1974. Pathogenesis of liver cancer. Archives of Pathology 98:145-148.

Firestone, D., Clower, M., Borsetti, A.P., Teske, R.H., and Long, P.E. 1979. Polychlorodibenzo-p-dioxin and pentachlorophenol residues in milk and blood of cows fed technical pentachlorophenol. Journal of Agricultural and Food Chemistry 27:1171-1177.

Fries, G.F. 1982. Potential polychlorinated biphenyl residues in animal products from application of contaminated sewage sludge to land. Journal of Environmental Quality 11:14.

Fries, G.F., and Jacobs, L.W. 1983. Tissue residues in livestock confined to unpaved lots with soil that contained polybrominated biphenyl. Journal of Animal Science 57(Supplement 1):302.

Fries, G.F., and Marrow, G.S. 1975. Excretion of polybrominated biphenyls in milk of cows. Journal of Dairy Science 58:947.

Fries, G.F., and Marrow, G.S. 1982. Soil ingestion by swine as a route of contaminant exposure. Environmental Toxicology and Chemistry 1:201-204.

Jones, G., and Butler, W.H. 1974. A morphological study of the liver lesion induced by 2,3,7,8-tetrachlorodibenzo-p-dioxin in rats. Journal of Pathology 112:93-97.

Kimbrough, R.D. (ed.) 1980. Halogenated biphenyls, terphenyls, naphthalenes, dibenzodioxins and related products. Topics in Environmental Health 4:406.

Kimbrough, R.D., Carter, C.D., Liddle, J.A., Cline R.E., and Phillips, P.E. 1977. Epidemiology and pathology of a tetrachlorodibenzodioxin poisoning episode. Archives of Environmental Health 32:77-86.

Kociba, R.J., Keyes, D.G., Beyer, J.E., Carreon, R.M., Wade, E.E., Dittenber, D.A., Kalnins, R.P., Frauson, L.F., Park, D.N., Barnard, S.D., Hummel, R.A., and Humiston, C.G. 1978. Results of a two year chronic toxicity and oncogenicity study of 2,3,7,8-tetrachlorodibenzo-p-dioxin (TCDD) in rats. Toxicology and Applied Pharmacology 46:279-303.

Krasovskii, G.N. 1976. Extrapolation of experimental data from animals to man. Environmental Health Perspectives 13:51-58.

Manis, J., and Kim. G. 1980. Polybrominated biphenyl: Acute and chronic effect on iron absorption and benzo(a)pyrene hydroxylase. Toxicology and Applied Pharmacology 54:41-47.

May, G. 1973. Chloracne from the accidental production of tetrachlorodibenzodioxin. British Journal of Industrial Medicine 30:276-283.

McConnell, E.E., Lucier G.W., Rumbaugh, R.C., Albro, P.W., Harvan, D.J., Hass, J.R., and Harris, M.W. 1984. Dioxin in soil: bioavailability after ingestion by rats and guinea pigs. Science 223:1077-1079.

Murray, F.J., Smith F.A., Nitschke, K.D., Humiston, C.G., Kociba, R.J., and Schwetz, B. A. 1979. Three generation reproduction study of rats given 2,3,7,8-tetrachlorodibenzo-p-dioxin (TCDD) in the diet. Toxicology and Applied Pharmacology 50:241-252.

National Research Council of Canada (NRCC). 1981. Polychlorinated dibenzo-p-dioxins. Publ. NRCC No. 18574 of the Environmental Secretariat. Ottawa, Canada: NRCC.

National Toxicology Program (NTP). 1982. Carcinogenesis bioassay of 2,3,7,8 tetrachlorodibenzo-p-dioxin (CAS Nol 1746-01-6) in Osborne-Mendel rats and B6C3Fl mice (gavage study). NTP Tech. Rep. Ser. Issue 209:195. NTP.

Office on Smoking and Health. 1982. The health consequences of smoking: Cancer. A report of the Surgeon General. Rockville, Maryland: Public Health Service, United States Department of Health and Human Services.

Poiger, H., and Schlatter, C. 1980. Influence of solvents and absorbents on dermal and intestinal absorption of TCDD. Food and Cosmetics Toxicology 18:477-481.

Stohs, S.J., Hassan, M.Q., and Murray, W.J. 1983. Lipid peroxidation as a possible cause of TCDD toxicity. Biochemical and Biophysical Research Communications 111:854-859.

Sweeney, G.D., and Jones, K.G. 1981. Studies of the mechanism of action of hepatotoxicity of 2,3,7,8-tetrachlorodibenzo-p-dioxin (TCDD) and related compounds. In: Human and Environmental Risks of Chlorinated Dioxins and Related Compounds, R.E. Tucker, A.L. Young, A.P. Gray, Eds., pp. 415-422. New York: Plenum Press.

Sweeney, G.D., Jones, K.G., Cole, F.M., Basford, D., and Krestynski, F. 1979. Iron deficiency prevents liver toxicity of 2,3,7,8-TCDD. Science 204:332-335.

U.S. EPA. 1980. Risk assessment on 2,4,5-trichlorophenoxyacetic acid (2,4,5-T), 2,4,5-trichlorophenoxypropionic acid (Silvex) and 2,3,7,8-tetrachlorodibenzo-p-dioxin (TCDD). EPA 600/6-81-003, NTIS PB81-234-825.

BACKGROUND EXPOSURE TO 2,3,7,8-TETRACHLORODIBENZO-p-DIOXIN

Perry J. Gehring, D.V.M, Ph.D.

Agricultural Products Research and Development,
and Health and Environmental Sciences U.S.A.
The Dow Chemical Company
Midland, Michigan 48640

This paper was submitted by Dr. Gehring after the symposium, to expand upon discussion that took place during the meeting.

From Public Health Risks of the Dioxins, proceedings of a symposium held on October 19-20, 1983 at The Rockefeller University, New York City. Edited by William W. Lowrance. Published by William Kaufmann, Los Altos, California.

It is well established that 2,3,7,8-tetrachlorodibenzo-p-dioxin (2,3,7,8-TCDD) is ubiquitous in the environment. Considerable effort is being made to ascertain what exposures may result from its presence in the environment at background levels as well as at levels exceeding background. Intrinsic in this effort is determining what fraction of 2,3,7,8-TCDD in various matrices may be absorbed (such as from skin exposure or from respiratory exposure to dust).

Although definitive answers to the absorption issues are elusive, it is feasible to estimate how much 2,3,7,8-TCDD is being absorbed by people exposed to background levels of the compound. This estimation is based on background levels of 2,3,7,8-TCDD in the fat of people with no known unique exposure to 2,3,7,8-TCDD. During this symposium it was reported that studies of three groups of such individuals revealed levels of 5-25 ppt.

Estimation of the level of 2,3,7,8-TCDD absorbed is based on the following assumptions:

- Liver and fat contain 50-fold greater levels of TCDD than other tissues. This assumption is consistent with the findings from studies on animals (Rose et al., 1976).
- The average human weighs 70 kg and has 10% body fat and a 1.75 kg liver.
- The half-life ($t_{½}$) for excretion in humans is 30 to 100 days. The $t_{½}$ for excretion in various other species varies from 11 to 30 days (Rose et al., 1976; Gasiewicz et al., 1979; Olson et al., 1980; Gasiewicz, 1983). For the following calculations $t_{½}$'s of 30, 60 and 100 days have been used.

Using 5 ppt (5 ng/kg) as the background level in fat, the total amount in the body may be calculated based on the preceding assumptions:

$$
\begin{aligned}
(7 \text{ kg fat})(5 \text{ ng/kg}) &= 35.00 \text{ ng} \\
(1.75 \text{ kg liver})(5 \text{ ng/kg}) &= 8.75 \\
(61.25 \text{ kg other tissue})(0.1 \text{ ng/kg}) &= \underline{6.12}
\end{aligned}
$$

TOTAL AMOUNT IN BODY 49.87 ng ($\simeq$ 50 ng)

It is reasonable to assume that the total amount of 2,3,7,8-TCDD in the body of individuals represents a steady state level (A_m^∞) acquired from repetitive exposure to background levels of the compound. A_m^∞ is attained when intake is equal to excretion. Assuming exposure occurs daily, the total dose absorbed daily (D_t) can be calculated from the equation (Gehring et al., 1976):

$$D_t = A_m^\infty/(1.44)(t_{1/2})$$

where for 5 ppt, $A_m^\infty = 50$ ng.

Table 1 shows the calculated D_t's and doses for several half-times.

Table 1. Background daily dose of 2,3,7,8-TCDD absorbed by individuals having 5 ppt in their fat.

	Daily Dose (D_t)**	
$t_{1/2}$/days*	ng/day	pg/kg/day
30	1.16	16.6
60	0.58	8.3
100	0.35	5.0

* $t_{1/2}$ is the half-life for clearance from the body.

** Total daily dose absorbed in body expressed as ng/day/70 kg person and pg/kg/day.

Of course, if the background level of 2,3,7,8-TCDD is greater than 5 ppt, the dose will also be proportionally greater.

This estimate of the total dose received by individuals exposed to background levels of 2,3,7,8-TCDD indicates that it is virtually impossible to reduce exposure to yet lower levels. Thus the rationale of such recommendations must be questioned.

REFERENCES

Gasiewicz, T.A., Geiger, L.E., Rucci, G., and Neal, R.A. Distribution, Excretion, and Metabolism of 2,3,7,8-Tetrachlorodibenzo-p-Dioxin in C57BL/6J, DBA/2J, and B6D2F$_1$/J Mice. Drug Metabolism and Disposition 11, (No. 5), 397-403 (1983).

Gasiewicz, T.A. and Neal, R.A. 2,3,7,8-Tetrachlorodibenzo-p-Dioxin Tissue Distribution, Excretion, and Effects on Clinical Chemical Parameters in Guinea Pigs. Toxicology and Applied Pharmacology 51, 329-339 (1979).

Gehring, P.J., Watanabe, P.G., and Blau, G.E. Pharmacokinetic Studies in Evaluation of the Toxicological and Environmental Hazard of Chemicals. New Concepts in Safety Evaluation, 21 (1976). Edited by M.A. Mehlman, R.E. Shapiro, and H. Blumenthal. Hemisphere Publishing Corporation, Washington, DC.

Olson, J.R., Gasiewicz, T.A., and Neal, R.A. Tissue Distribution, Excretion, and Metabolism of 2,3,7,8-Tetrachlorodibenzo-p-Dioxin (TCDD) in the Golden Syrian Hamster. Toxicology and Applied Pharmacology 56, 78-85 (1980).

Rose, J.Q., Ramsey, J.C. Hummel, R.A., and Gehring P.J. The Fate of 2,3,7,8-Tetrachlorodibenzo-p-Dioxin (TCDD) Following Single and Repeated Oral Administration in Rats. Toxicology and Applied Pharmacology 36, 339-352 (1976).

DIOXINS AS CARCINOGENIC PROMOTERS

I. Bernard Weinstein, M.D.

Division of Environmental Sciences
School of Public Health, and Cancer Center/
Institute of Cancer Research
Columbia University
New York, New York 10032

This paper summarizes Dr. Weinstein's presentation during the symposium. It is adapted, with permission, from a Letter to the Editor, Science 219:794-796 (1983), copyright 1983 by the AAAS.

From Public Health Risks of the Dioxins, proceedings of a symposium held on October 19-20, 1983 at The Rockefeller University, New York City. Edited by William W. Lowrance. Copyright, The Rockefeller University, 1984. Published by William Kaufmann, Los Altos, California.

In a multistage liver carcinogen assay, TCDD has been shown to act as a "tumor promoter" (1). In initiation--promotion tests, a low dose of a chemical agent, called the "initiator", is administered that does not (at that dose) induce tumors; then repeated applications are made of the chemical being tested, and if this results in tumors, the latter agent is designated a "promoter". Many promoters, including TCDD, seem not to act on the genetic apparatus of cells directly, but exert their effect "epigenetically". Thus, although they do not damage DNA directly, they can affect cellular growth and differentiation, and alter a number of properties of cell membranes. Neither the biological significance of these effects nor their implications for environmental regulation is clear. Several steps almost certainly are involved in all carcinogenic processes. Further studies on TCDD may provide insights into these questions.

For the moment, I am extremely skeptical about certain proposals that would establish separate guidelines for evaluating the health effects of "genotoxic" and "epigenetic" carcinogens for regulatory purposes, which would tend to soften the restrictions on the latter class of agents.

We are not yet able precisely to distinguish between those carcinogens that act through genotoxic mechanisms and those that act through epigenetic mechanisms, or to know the implications that the laboratory "promoter" tests have for whole organisms, especially humans. Indeed, recent studies in molecular genetics, developmental biology, and immunology tend to blur the classical distinctions between genetic and epigenetic mechanisms, even in normal biologic processes (2,8).

Even if this distinction were valid, our current methods for assessing whether or not a given agent is likely to be genotoxic in humans have very serious limitations (3). Worse, at present we do not have well-validated short-term tests for assessing agents that act through nongenotoxic mechanisms, that is, tumor promoters, hormones, and so forth. Identification of "epigenetic" agents must, therefore, often be done by exclusion, a risky approach.

Most of the known carcinogens produce multiple effects. In fact, when given at sufficient dosage the genotoxic chemicals are usually complete carcinogens and, therefore, probably produce both tumor-initiating and tumor-promoting effects (4,5). Simple tests for genotoxicity may fail to assess the promoting capacity of these compounds. This and other factors severely limit attempts to predict the mechanism(s) of action and relative potencies of carcinogens, when findings based simply on genotoxic activity are used. The paradigm of random point mutation as a basis for understanding the carcinogenic action of agents that display genotoxic effects may itself be antiquated, in view of the multistage aspects of the carcinogenic process, probable synergistic (and sometimes inhibitory) multifactor interactions, and the possibility that carcinogenesis involves more complex genomic changes (gene rearrangements, chromosomal translocation, oncogene activation, altered DNA methylation, and so forth) (2).

Certain tumor promoters (such as the phorbol esters and TCDD) can induce a significant number of tumors in animals, even without prior application of an initiating carcinogen (4). In addition, there are a few studies suggesting that, although the primary target of the phorbol ester tumor promoters is cellular membranes rather than DNA (2), these compounds may indirectly inflict chromosomal damage, perhaps via the generation of activated forms of oxygen (6). If this is the case, then these compounds also have genotoxic activity, albeit through an indirect effect.

It is often assumed that tumor promoters and other agents that might act through epigenetic mechanisms will, in contrast to initiating and genotoxic carcinogens, display a threshold in their dose response. The data on dose-response relationships with tumor promoters are skimpy, and I know of no evidence that clearly establishes a threshold for tumor promoters in humans or in experimental systems. Even if this were the case, how would we know how to extrapolate from a specific set of data the actual threshold level in a heterogeneous human population?

It is true that the known tumor promoters require repeated application to exert their tumor-producing effect, whereas the single application of certain initiating carcinogens is sufficient (4). This does not necessarily imply a comfortable margin of safety for tumor promoters, because for many substances that are of concern (such as water pollutants, industrial chemicals, and food additives) there is likely to be repeated and prolonged human exposure. Moreover, some of these substances are only slowly degraded and, therefore, will persist or even accumulate in body tissues or the general environment.

There is the impression that tumor promoters are much less potent than initiating carcinogens and, therefore, are less hazardous. This is not necessarily the case. On a molar basis TPA is about two orders of magnitude more potent in exerting biologic effects than benzo[a]pyrene (4), and TCDD is one of the most potent of the promoters.

We know that nature has evolved specific defense mechanisms against some of the genotoxic agents, including conjugation and detoxifying mechanisms and DNA excision repair. We do not know to what extent humans have evolved protective mechanisms against tumor promoters. I do not doubt that such mechanisms exist, but at the present time we do not know their properties or relative efficiencies.

A final reason for being concerned about the potential health hazards of tumor promoters and various carcinogenic cofactors that do not appear to act by directly damaging cellular DNA is the evidence that a major fraction of human cancer is due to "lifestyle factors" and that many of these may not act as simple genotoxic agents (7). It is essential, therefore, that we not overemphasize our concern with genotoxic agents, downplay the potential health hazards of other types of agents, and thus distort priorities in our efforts at primary cancer prevention.

In summary, although there has been exciting progress in our understanding of the mechanism of action of environmental carcinogens, the field is in sufficient flux that at present it would be premature to alter the existing, well-established guidelines for risk extrapolations of potential hazards to humans.

Especially, I see no justification for assuming a nonlinear dose response and threshold model for certain carcinogens simply because they do not give a positive response in certain currently used assays for genotoxicity.

REFERENCES

1. Pitot, H.C., Goldsworthy, T., Campbell, H.A., and Poland, A. Cancer Research 40:3616 (1980).

2. Weinstein, I.B. Journal of Supramolecular Structure and Cellular Biochemistry 17:99 (1981).

3. Fisher, P.B. and Weinstein, I.B. In: Carcinogens in Industry and Environment (J.M. Sontag, Ed.), Dekker, New York, 1981, p.113; Hollstein, M, McCann, J., Angelosanto, F.A., and Nichols, W.W. Mutation Research 65:13 (1979).

4. Slaga, T.J., Sivak, A., and Boutwell, R.K., Editors, Carcinogenesis, Vol.21 Mechanisms of Tumor Promotion and Cocarcinogenesis, Raven, New York, 1978.

5. Ivanovic, V. and Weinstein, I.B. Carcinogenesis 3:505 (1981).

6. Kinsella, A.R. and Radman, M. Proceedings of the National Academy of Sciences U.S.A. 75:6149 (1978); Nagasawa, H. and Little, J.B. Ibid. 76:1943 (1979); Birnboim, H.C. Science 215:1247 (1982).

7. Doll, R. and Peto, R. The Causes of Cancer, Oxford University Press, Oxford, 1981.

8. Weinstein, I.B., Horowitz, A., Jeffrey, A., and Ivanovic, V. In: Genes and Proteins in Oncogenesis (I.B. Weinstein and H.J. Vogel, Eds.), Academic Press, New York, 1983, p.99.

MODIFICATION OF CHEMICAL CARCINOGENESIS BY HALOGENATED HYDROCARBONS AND OTHER ENZYME INDUCERS

John DiGiovanni, Ph.D.

The University of Texas System Cancer Center
Smithville, Texas 78735

From Public Health Risks of the Dioxins, proceedings of a symposium held on October 19-20, 1983 at The Rockefeller University, New York City. Edited by William W. Lowrance. Published by William Kaufmann, Los Altos, California.

Many compounds of diverse chemical structure possess the ability to induce microsomal monooxygenase enzymes (1,2). In addition, many of these compounds affect the activities of other microsomal and soluble enzymes as well. Enzyme induction has been postulated as a mechanism for anticarcinogenesis by a number of compounds, some of which are listed in Table 1.

Table 1. Inhibition of chemical carcinogenesis by halogenated hydrocarbons and related enzyme inducers.[a,b]

Inducer	Carcinogen	Species	Tissue
1. Aroclor 1254	DMBA	mouse	skin
	AFB_1	trout	liver
2. Kanechlor 500	3'-Me-DAB	rat	liver
	AAF	rat	liver
	DEN	rat	liver
3. Kanechlor 400	3'-Me-DAB	rat	liver
4. 3,4,3',4'-TCB	DMBA	mouse	skin
5. 2,3,7,8-TCDD	DMBA,MCA,B[a]P, B[a]P-diol-epoxide, 7-MBA,12-MBA,5-MeC, DB[a,h]A	mouse	skin
6. DDT	DMBA	rat	mammary gland
7. 4-Methyl-coumarin	DMBA	rat	mammary gland
8. MCA	3'-Me-DAB	rat	liver
	4'-Fl-DAB[a]	rat	liver
	2',4'-diFl-DAB	rat	liver
	AAF	rat	liver
	7-Fl-AAF	rat	liver
9. Pb	AFB_1	rat	liver
	AAF	rat	liver
	urethan	mouse	lung
10. Indole-3-carbinol	DMBA	rat	mammary gland
	B[a]P	mouse	forestomach
3',3-di-indolymethane	DMBA	rat	mammary gland
	B[a]P	mouse	forestomach
Indole-3 acetonitrile	B[a]P	mouse	forestomach
11. Phenothiazines			
Phenothiazine	DMBA	rat	mammary gland
Chlorpromazine	DMBA	rat	mammary gland

Table 1. Continued.

[a] DDT, 1,1,1-trichloro-2,2-bis-(p-chlorophenyl)ethane; 4'-Fl-DAB,4'-Fluoro-4-dimethylaminoazobenzene; 2'4'-diFl-DAB, 2',4',-difluoro-4-dimethylaminoazobenzene; 7-Fl-AAF, 7-fluoro-acetylaminofluorene.

[b] Data taken from Slaga, T.J. and DiGiovanni, J. Inhibition of Chemical Carcinogenesis, American Chemical Society Monograph (in press).

In addition, this mechanism for anticarcinogenesis appears to apply to certain flavonoids (e.g., 5,6-benzoflavone) which have been widely studied (reviewed in 3).

The most potent member and prototype of this group of inhibitors is 2,3,7,8-tetrachlorodibenzo-p-dioxin (TCDD). Figure 1 illustrates the dose-response relationship for inhibition of 7,12-dimethylbenz-[a]anthracene (DMBA) skin tumor-initiation by TCDD in SENCAR mice (4).

Figure 1. Dose-response for the anticarcinogenic effect of TCDD on tumor initiation by DMBA. Groups of 30 female SENCAR mice were treated topically with various doses of TCDD 3 days prior to initiation with 10 nmol DMBA. One week after initiation, mice received twice weekly applications of 3.4 nmol TPA. The ED_{50} (defined here as the dose of TCDD producing half-maximal inhibition) is 5.6 pmol per mouse.

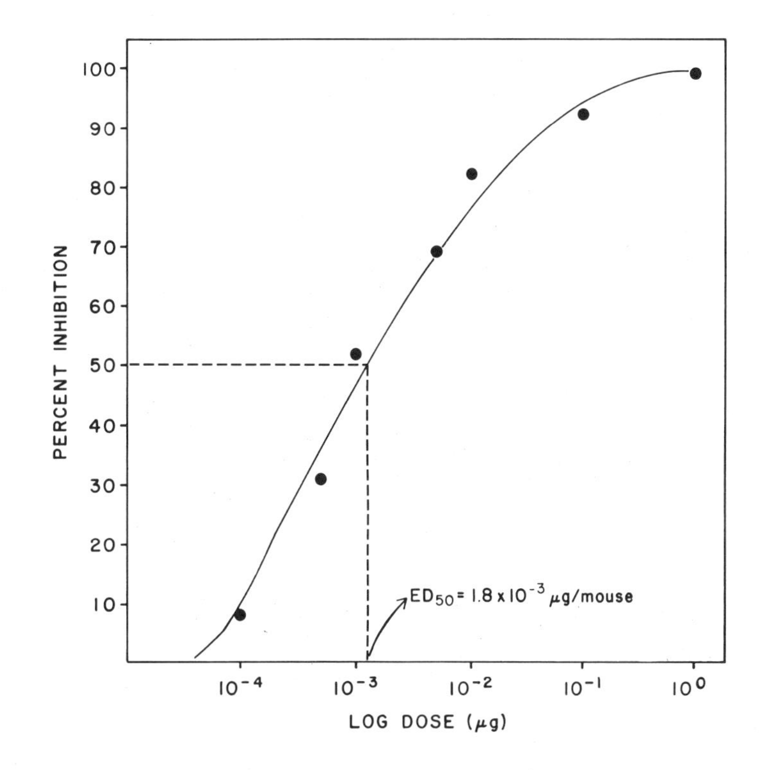

In these experiments, TCDD was applied topically at various doses 3 days prior to initiation with 10 nmol DMBA. One week after initiation, mice received twice weekly applications of 12-0-tetra-decanoylphorbol-13-acetate (TPA) for 20 weeks. The dose of TCDD producing a half-maximal inhibition after 20 weeks of promotion with TPA was 5.6 pmol per mouse (1.8×10^{-3} ug/mouse).

TCDD is a highly toxic and widely studied chemical found as an environmental contaminant (reviewed in 5 and 6). Like the other compounds listed in Table 1, TCDD is known to markedly influence enzyme pathways responsible for both the activation and inactivation of polycyclic aromatic hydrocarbon (PAH) carcinogens as well as other chemical carcinogens. TCDD is an extremely potent inducer of hepatic microsomal monooxygenase activity with properties similar to that of 3-methylcholanthrene (MCA)(7). Microsomal enzyme induction by MCA is characterized by increased cytochrome(s) P_1-450, a spectrally distinct cytochrome(s) and a shift to a narrow substrate specificity (1,8,9). TCDD has the added distinctions of being 30,000 times more potent than MCA and markedly affecting other enzyme pathways as noted above (7,10-13). When mice are treated with TCDD, the time course and magnitude for inhibition of mouse skin tumor initiation by various PAH follows closely the time course and magnitude for the induction of aryl hydrocarbon hydroxylase (AHH, E.C. 1.14.14.2) and uridine-5'-diphosphoglucuronyltransferase (UDPGT) in skin (14-17). Further studies from our laboratories have shed light on the mechanism by which TCDD inhibits skin carcinogenesis with PAH (18,19) (Table 2).

Table 2. Effect of pretreatment with TCDD on covalent binding of ^{3}H-DMBA to epidermal macromolecules.[a]

		Hydrocarbon Bound to Macromolecules (pmol/mg x10^2)		
Treatment		DNA	RNA	Protein
DMBA:	3 hr[b]	4.79	2.88	69.5
	24 hr	12.63	7.16	38.8
DMBA + TCDD[c]:	3 hr	1.78	1.56	63.8
	24 hr	3.52	3.16	39.4

Table 2. Continued.

[a] Mice were treated topically with 10 nmol of ^{3}H-DMBA (10 uCi) and sacrificed 3 and 24 hr later. Data taken from reference 18.

[b] Time of sacrifice after application of ^{3}H-DMBA; 40 mice were used for each experimental group.

[c] Mice in this group received 1 ug of TCDD (topically) 72 hr prior to application of ^{3}H-DMBA.

In these experiments, TCDD was applied topically at doses and times producing maximal or near maximal inhibition of tumor formation. Under these conditions, a marked reduction in the binding of ^{3}H-DMBA to epidermal DNA and RNA was observed. These results suggest that under the influence of TCDD, epidermal cells were programmed to inactivate PAH carcinogens more efficiently.

Several pure polychlorinated biphenyl (PCB) isomers have been tested for their antagonistic effects on chemical carcinogenesis (Table 3).

Table 3. Effects of TCDD and congeners of TCDD on the tumor-initiating activity of DMBA and epidermal AHH and UDPGT activities.

Thirty female CD-1 mice were used per experimental group. All mice were initiated with 10 nmol DMBA. Animals were pretreated with single topical doses 3 days prior to initiation as follows: TCDD 1 ug/mouse; DCDD, 100 ug/mouse; TCB, 625 ug/mouse; HCB, 625 ug/mouse. One week after initiation, mice received twice weekly applications of 17 nmol TPA for 24 weeks. Control animals receiving the acetone vehicle followed by TPA promotion had 0.03 papillomas/mouse.

	Effect on DMBA tumor initiation and enzyme activity			
Compound	Papillomas/ mouse	% of control[a]	AHH[b]	UDPGT[c]
TCDD	0.34	9.0	954	0.935
TCB	0.40	10	702	0.840
HCB	3.36	88	50	0.420
DCDD	3.44	91	41	0.430

Table 3. Continued.

[a] Average number of papillomas per mouse expressed as percentage of the DMBA-initiated, TPA-promoted group (3.8 papillomas/mouse). Data taken from reference 16.

[b] Specific activities are expressed as pmol 3-hydroxybenzo(a)pyrene formed per mg protein per min of incubation. Specific activity using epidermal microsomes from control mice was 45.

[c] Specific activities are expressed as nmol p-nitrophenyl glucuronide formed per mg microsomal protein per min of incubation. Specific activity using epidermal microsomes from control mice was 0.410.

In the two-stage system of mouse skin tumorigenesis, Aroclor 1254 (a PCB mixture) and 3,4,3',4'-tetrachlorobiphenyl (TCB) inhibited DMBA tumor initiation (16,17). TCB is a pure PCB isomer possessing MCA-type inducing properties (20,21). When 2,4,5,2',4',5'-hexachlorobiphenyl (HCB) was tested, it had no effect on tumor initiation by DMBA (16,19). HCB is another pure PCB isomer but possesses phenobarbital (Pb)-type inducing properties (20,21). Induction of microsomal monooxygenase enzymes with Pb is characterized by an increased concentration of cytochrome(s) P-450. Furthermore, Pb enhances the metabolism of a wide variety of substrates (1-5). Apparently, MCA-type inducing activity is a prerequisite for inhibiting PAH tumor-initiation on mouse skin. This does not appear to be the case, however, for inhibition of hepatic tumors induced by other classes of carcinogens (22,23,24). For example, Kanechlor 500 effectively inhibited the formation of hepatic tumors induced by 3'-methyl-4-dimethylaminoazobenzene (3'-Me-DAB), 2-acetylaminofluorene (AAF), and diethylnitrosamine (DEN) (25). Kimura et al. (26) reported that inclusion of Kanechlor 400 in the diets of rats 4 months prior to and 2 months during 3'-Me-DAB administration completely protected against the formation of hepatocarcinomas induced by this carcinogen. Kanechlor 500 and Kanechlor 400 are PCB mixtures containing isomers with both MCA and Pb-type inducing properties (20,21,27,28). MCA (29) and Pb (24) both effectively inhibited the induction of hepatic tumors by AAF.

The time of treatment and sequence of exposure to the chemical modifier and carcinogen were extremely critical in all of the studies cited above. When Kanechlor 400 was administered subsequent to 3'-Me-DAB, the incidence of hepatocarcinomas was significantly increased (26). This finding as well as other reports suggest that PCBs are promoters of hepatic tumors when applied after exposure to the carcinogen (27,30,31). Pb is known to enhance the development of hepatocarcinomas induced in the livers of rats following sequential partial hepatectomy and AAF administration (24,32). Recent studies by Kouri et al. (33,34) regarding the effects of subcutaneous coinjections of TCDD on MCA-induced fibrosarcomas suggested a potential cocarcinogenic effect in certain mouse strains (e.g., DBA/2 mice). Furthermore, Pitot et al. (35) demonstrated that TCDD was a potent promoting agent for hepatocarcinogenesis in rats exposed to DEN. Finally, a recent report by Poland et al. (36) demonstrated that TCDD was capable of promoting skin tumors in previously initiated HRS/J hairless mice but not in haired mice. Earlier reports from our laboratories failed to demonstrate tumor initiating or promoting properties of TCDD in haired mice such as CD-1 and SENCAR (37,38).

In summary, the chlorinated hydrocarbons and other related enzyme inducers can be seen as a double-edged sword. On the one hand, exposure to these agents prior to a carcinogen results in an anticarcinogenic effect. The mechanistic studies that have been conducted indicate that antagonism of tumorigenesis correlates with increased enzyme activities (both oxidative and conjugative) and a reduction in the formation of DNA-adducts in the target tissue. The formation of critical electrophilic intermediates from various carcinogens depends on a balance between their rate of formation and rate of removal. Under conditions of prior exposure to enzyme inducing agents such as TCDD, PCBs, Pb, etc., this balance is apparently shifted toward detoxification.

On the other hand, under certain conditions, prior exposure to enzyme inducers may enhance tumor formation. In certain mouse strains, basal levels of AHH are low (39,40). Several reports have suggested that enzyme inducers including 5,6-BF and TCDD enhance PAH carcinogenesis in these strains by elevating AHH levels and shifting the balance of metabolism toward greater activation (33,34,41).

In addition, many of the halogenated hydrocarbons appear to possess tumor-promoting activity when administered after exposure to carcinogens. Thus, although inducers provide protection from carcinogens under certain conditions using particular experimental protocols, deviations from these protocols can have serious, in fact, opposite consequences. It may, therefore, be very difficult to determine the outcome of exposure to these agents when exposure to other carcinogenic agents occurs at about the same time.

REFERENCES

1. Conney, A.H., Pharmacological Reviews (1967) 9, 37.

2. Remmer, H., European Journal of Clinical Pharmacology (1972) 5, 166.

3. Slaga, T.J. (ed.), Modifiers of Chemical Carcinogenesis, An Approach to the Biochemical Mechanism and Cancer Prevention, Raven Press, New York, 1980.

4. DiGiovanni, J., Decina, P.C., and Diamond, L., Carcinogenesis (1983) 4, 1045.

5. Poland, A., and Kende, A., Federation Proceedings (1976) 35, 2404.

6. Poland, A., and Knutson, J.C., Annual Review of Pharmacology and Toxicology (1982) 22, 517.

7. Poland, A., and Glover, E., Molecular Pharmacology (1974) 10, 349.

8. Conney, A.H., In: Fundamentals of Drug Metabolism and Drug Disposition (B.N. La Du, H.G. Mandel, and E.L. Way, eds.), Williams and Wilkins, Baltimore, 253, 1972.

9. Sladek, N.E., and Mannering, G.J., Biochemical and Biophysical Research Communications (1966) 23, 668.

10. Owens, I.S., Journal of Biological Chemistry (1977) 252, 2827.

11. Baars, A.J., Jansen, M., and Breimer, D.D., Biochemical Pharmacology (1978) 27, 2487.

12. Fowler, B.A., Hook, G.E.R., and Lucier, G.W., Journal of Pharmacology and Experimental Therapeutics (1977) 203, 712.

13. Lucier, G.W., McDaniel, O.S., and Hook, G.E.R., Biochemical Pharmacology (1975) 24, 325.

14. Berry, D.L., Slaga, T.J., DiGiovanni, J., and Juchau, M.R., Annals of the New York Academy of Sciences (1979) 320, 405.

15. Cohen, G.M., Bracken, W.M., Iyer, P.R., Berry, D.L., and Slaga, T.J., Cancer Research (1979) 39, 4027.

16. DiGiovanni, J., Berry, D.L., Gleason, G.L, Kishore, G.S., and Slaga, T.J., Cancer Research (1980) 40, 1550.

17. DiGiovanni, J., Kishore, G.S., Slaga, T.J., and Boutwell, R.K., In: Polynuclear Aromatic Hydrocarbons: Chemistry and Biological Effects (A. Bjorseth and A.J. Dennis, eds.), Battelle Press, Columbus, Ohio, 935, 1983.

18. DiGiovanni, J., Berry, D.L., Juchau, M.R., and Slaga, T.J., Biochemical and Biophysical Research Communications (1979) 86, 577.

19. DiGiovanni, J., Berry, D.L., Slaga, T.J., and Juchau, M.R., In: Polynuclear Aromatic Hydrocarbons (D.W. Jones and P. Leber, eds.), Ann Arbor Science Publishers, Inc., Ann Arbor, Michigan, 553, 1979.

20. Goldstein, J.A., Hickman, P., Bergman, H., McKinney, J.D., and Walker, M.P., Chemico-Biological Interactions (1977) 17, 69.

21. Poland, A., and Glover, E., Molecular Pharmacology (1977) 13, 924.

22. Yamamoto, R.S., Weisburger, J.H., and Weisburger, E.K., Cancer Research (1971) 31, 483.

23. McLean, A.E.M., and Marshall, A., British Journal of Experimental Pathology (1971) 52, 322.

24. Peraino, C., Fry, R.J.M., and Staffeldt, E., Cancer Research (1971) 31, 1506.

25. Makiura, S., Aoe, H., Sugihara, S., Hirao, K., Arai, M., and Ito, N., Journal of The National Cancer Institute (1974) 53, 1253.

26. Kimura, N.T., Kanematsu, T., and Baba, T., Zeitschrift fur Krebforschung (1976) 87, 257.

27. Ito, N., Nagasaki, H., Arai, M., Makiura, S., Sugihara, S., and Hirao, K., Journal of The National Cancer Institute (1973) 51, 1637.

28. Yoshimura, H., Ozawa, N., and Saeki, S., Chemical and Pharmaceutical Bulletin (1978) 26, 1215.

29. Miller, E.C., Miller, J.A., Brown, R.R., and MacDonald, J.E., Cancer Research (1958) 18, 469.

30. Nishizumi, M., Cancer Letters (1976) 2, 11.

31. Preston, B.D., Van Miller, J.P., Moore, R.W., and Allen, J.R., Journal of The National Cancer Institute (1981) 66, 509.

32. Peraino, C., Fry, R.J.M., and Grube, D.D., In: Carcinogenesis, Mechanisms of Tumor Promotion and CoCarcinogenesis, Vol. II (T.J. Slaga, A. Sivak and R.K. Boutwell, eds.), Raven Press, New York, 421, 1978.

33. Kouri, R.E., In: Polynuclear Aromatic Hydrocarbons: Chemistry, Metabolism and Carcinogenesis, Vol. 1, Carcinogenesis: A Comprehensive Survey, Raven Press, New York, 137, 1976.

34. Kouri, R.E., Rude, T.H., Joglekar, R., Dansette, P.M., Jerina, D.M., Atlas, S.A., Owens, I.S., and Nebert, D.W., Cancer Research (1978) 38, 1977.

35. Pitot, H.C., Goldsworthy, T., Campbell, H.A., and Poland, A., Cancer Research (1980) 40, 3616.

36. Poland, A., Palen, D., and Glover, E., Nature (1982) 300, 271.

37. Berry, D.L., DiGiovanni, J., Juchau, M.R., Bracken, W.M., Gleason, G.L., and Slaga, T.J., Research Communications in Chemical Pathology and Pharmacology (1978) 20, 101.

38. DiGiovanni, J., Viaje, A., Berry, D.L., Slaga, T.J., and Juchau, M.R., Bulletin of Environmental Contamination and Toxicology (1977) 18, 552.

39. Nebert, D.W., Robinson, J.R., Niwa, A., Kimaki, K., and Poland, A.P., Journal of Cellular Physiology (1975) 85, 393.

40. Robinson, J.R., Considine, N., and Nebert, D.W., Journal of Biological Chemistry (1974) 249, 5851.

41. Lesca, P., Carcinogenesis (1981) 2, 199.

CLASSIFICATION AND PATHOLOGICAL DIAGNOSIS OF SOFT TISSUE SARCOMAS

Steven I. Hajdu, M.D.

Attending Pathologist and Chief of the Cytology Service
Memorial Sloan-Kettering Cancer Center
New York, New York 10021

From Public Health Risks of the Dioxins, proceedings of a symposium held on October 19-20, 1983 at The Rockefeller University, New York City. Edited by William W. Lowrance. Copyright, The Rockefeller University, 1984. Published by William Kaufmann, Los Altos, California.

OVERVIEW OF SOFT TISSUE TUMORS

Soft tissue sarcomas arise from derivatives of the embryonic mesoderm. The embryonic mesoderm is segmented into a series of somites and split into somatic and splanchnic layers that are the source of the mesenchymal or connective tissues (1,2). Mesenchymal cells in the course of differentiation become specialized and may assume the cytologic characteristic of fibroblasts, myoblasts, lipoblasts, chondroblasts, osteoblasts or an endless number of other primitive forms. The fact that mesenchymal cells are ubiquitous and highly versatile is further complicated, since any well-defined mesenchymal cell may either undergo maturation arrest, so-called dedifferentiation, or at the conclusion of the mitotic cycle reach a higher level of differentiation by acquiring complex cytoplasmic organelles, depositing biochemically and immunologically active products, and assuming a different phenotype from that of the parent cell (1-4).

Soft tissues are composed of clones of mesenchymal cells such as myoblasts, adipocytes and fibroblasts. Both differentiated and undifferentiated tumors may be composed of a combination of cellular elements at various stages of differentiation and may grow in a variety of tissue patterns. Once it is recognized that cell morphology and tissue patterns are subject to modulation and are influenced by local tissue conditions and a host of other factors, it is not difficult to understand that mesenchymal tissues and tumors may assume (permanently or temporarily) dangerously misleading and overlapping tissue patterns and cell morphology. Table 1 lists some of the descriptive characteristics that must be considered.

Table 1. Some descriptive characteristics that must be considered in classifying sarcomas.

GROWTH PATTERN	CELL MORPHOLOGY
Arranged	Slender spindle
Spreading	Plump spindle
Lacy	Granular epithelioid
Epithelioid	Clear epithelioid
Alveolar	Isomorphic giant
Disarranged	Pleomorphic giant

Table 1. Continued.

STROMA APPEARANCE	CELL PRODUCTS
Fibrillar	Collagen
Sclerosed	Glycogen
Myxoid	Polysaccharides
Vascular	Fat
Inflamed	Melanin
Necrotic	Secretory granules
Chondrified	Crystals
Ossified	Fine structure
Calcified	Tissue antigens

Modified from Hajdu (2).

Therefore, one should not be alarmed that soft tissue neoplasms are commonly labelled differently by different pathologists and occasionally by the same pathologist in limited tissue samples. In soft tissue tumors, the relationship between differentiation and cell of origin is often blurred by the non-specific assembly of undifferentiated cellular elements that may show no apparent structural differences between reactive and neoplastic growths. In 1919, James Ewing wrote that "the capacity of connective tissue to indulge in exuberant reactive or reparative growth is remarkable." No one would deny that actively growing reparative tissue (e.g., granulation tissue), fasciitis, or myositis ossificans may contain all, or nearly all, the cellular elements that embryonic mesenchyme can produce and may mimic neoplastic growth.

The cell of origin of many soft tissue tumors is far from settled, but there is a general agreement among investigators that the tumors derive from primitive pluripotential mesenchymal cells (2,5). There is no statistically reliable figure for soft tissue tumors, but it is estimated that over 6,000 new cases of soft tissue sarcomas are diagnosed annually in the United States (2,3). This represents less than one percent of all malignant neoplasms in adults and seven percent in children. Because of their rarity, soft tissue sarcomas represent a minor part of the diagnostic experience of pathologists. There are about two dozen centers worldwide that are large enough to see a sufficient number of soft tissue tumors to be familiar with the variants. There are over 190 well-defined microscopic forms of soft tissue tumors, of which 67 are malignant neoplasms, 61 are benign neoplasms, and 70 are reactive, non-neoplastic lesions that may resemble neoplasms (Table 2). Also, benign neoplastic and non-neoplastic lesions outnumber malignant neoplasms by about a hundred to one. Due to overlapping morphology of many benign and malignant lesions and inability of pathologists to recognize "borderline connective tissue lesions" and "sarcoma in situ", connective tissue neoplasms traditionally are called either benign or malignant.

Table 2. Classifications of soft tissue tumors and tumor-like lesions.

Non-Neoplastic Lesions	Benign Neoplastic Lesions	Malignant Neoplastic Lesions
FIBROUS TISSUE TUMORS		
Fasciitis	Benign fibroblastic fibrous histiocytoma	Malignant fibroblastic fibrous histiocytoma
Keloid	Benign histiocytic fibrous histiocytoma	Malignant histiocytic fibrous histiocytoma
Elastofibroma	Benign pleomorphic fibrous histiocytoma	Malignant pleomorphic fibrous histiocytoma
Fibromatosis	Fibroma	Desmoid tumor
		Fibroblastic fibrosarcoma
		Pleomorphic fibrosarcoma
TENDOSYNOVIAL TISSUE TUMORS		
Tendosynovial cyst	Fibroma	Malignant giant cell tumor
Tendosynovitis	Lipoma	Biphasic tendosynovial sarcoma
Synovial chondromatosis	Villonodular synovitis	Monophasic spindle cell type tendosynovial sarcoma
	Benign giant cell tumor	Monophasic pseudoglandular type tendosynovial sarcoma
		Epithelioid sarcoma
		Clear cell sarcoma
		Chordoid sarcoma
ADIPOSE TISSUE TUMORS		
Fat necrosis	Well-differentiated lipoma	Well-differentiated liposarcoma
Lipogranuloma	Myxoid lipoma	Myxoid liposarcoma
Lipodystrophia	Fibroblastic lipoma	Lipoblastic liposarcoma
Piezogenic papules	Pleomorphic lipoma	Fibroblastic liposarcoma
Adiposis dolorosa	Angiolipoma	Pleomorphic liposarcoma
Steatopygia	Angiomyolipoma	
	Myelolipoma	
	Lipoblastoma	
	Hibernoma	

Table 2. Continued.

Non-Neoplastic Lesions	Benign Neoplastic Lesions	Malignant Neoplastic Lesions
MUSCLE TUMORS		
Atrophy Dystrophy Polymyositis Rhabdomyolysis Proliferative myositis Fibromatosis Myositis ossificans	Leiomyoma Angiomyolipoma Rhabdomyoma	Leiomyosarcoma Leiomyoblastoma Embryonal rhabdomyosarcoma Pleomorphic rhabdomyosarcoma Rhabdomyoblastoma
VESSEL TUMORS		
Pyogenic granuloma Angioblastic lymphoid hyperplasia Hereditary hemorrhagic telangiectasia Arteriovenous fistula	Capillary hemangioma Cavernous hemangioma Venous hemangioma Papillary endothelial hyperplasia Hypertrophic hemangioma Angiomatosis Angiofibroma Angiolipoma Angiomyolipoma Hemangioblastoma Glomus tumor Lymphangioma Lymphangiomyoma	Hemangiopericytoma Hemangiosarcoma Kaposi's sarcoma Lymphangiosarcoma
PERIPHERAL NERVE TUMORS		
Degeneration Hypertrophy Traumatic neuroma	Neurofibroma Neurofibromatosis Angioneuroma Plexiform neuroma Pacinian neurofibroma Glandular schwannoma Pleomorphic schwannoma Cystic schwannoma Storiform schwannoma Epithelioid schwannoma	Malignant collagenous schwannoma Malignant non-collagenous schwannoma Primitive neuroectrodermal tumor
EXTRASKELETAL BONE TUMORS		
Chondroid metaplasia Osteoid metaplasia Myositis ossificans	Chondroma Osteochondroma Osteoma	Osteogenic sarcoma Chondrosarcoma Ewing's sarcoma

Table 2. Continued.

Non-Neoplastic Lesions	Benign Neoplastic Lesions	Malignant Neoplastic Lesions
MISCELLANEOUS SOFT TISSUE TUMORS		
	Benign granular cell tumor	Malignant granular cell tumor
	Benign mesenchymoma	Alveolar soft part sarcoma
		Malignant lymphoma
		Granulocytic sarcoma
		Plasmacytoma
		Postirradiation sarcoma
		Malignant mesenchymoma
		Undifferentiated soft tissue sarcoma

Modified from Hajdu (1).

Numerous problems about the histogenesis and pathology of soft tissue sarcomas remain. Whatever the histogenesis, pathologists may enhance the diagnosis of soft tissue neoplasms by knowledge of the clinical presentation, size, site, radiologic appearance, and age of the patient. Needless to say, close cooperation among the surgeon, radiologist, and pathologist is essential for arriving at an accurate microscopic diagnosis. The role of the pathologist in the histologic diagnosis of soft tissue sarcomas is crucial; it is a difficult and complex task with serious therapeutic ramifications (1,2,5). As William Boyd said, "It is the high function of the pathologist not merely to attach correct labels to lesions, but to reconstruct the course of events from the earliest inception of disease to the final moment of life."

The understanding of soft tissue sarcomas was held back for centuries by misuse and misunderstanding of various names and definitions, inaccurate information, or plain ignorance. Galen (A.D. 130-201) defined sarcomas in such a way that many forms of inflammation or infectious swelling as well as benign neoplasms were called sarcoma. As late as the mid-1800's, Carl Rokitansky wrote that "sarcomata represent benign new growths, they are always purely local affections, they are curable by complete extirpation: that is they do not recur at the same spot, and still less do they multiply in other localities" (1). Current classification schemes of soft tissue tumors separate tumors whose cell of origin is presumed to be known from those whose histogenesis is not known. As new diagnostic techniques develop, it will be possible to establish the histogenesis of soft tissue tumors even more accurately. Histogenetic classification, based on characteristic microscopic features, permits subdivision of soft tissue tumors regardless of the anatomic site of the tumor or age of the patient (see Table 2). Although histogenesis may remain unsettled and controversial in many instances, it should nonetheless be included in therapeutic planning (1,2).

Every soft tissue sarcoma like any other malignant neoplasm must have a beginning and evolve from the preneoplastic and incipient phases to the established and invasive phases. The rate of progression from one phase to another may vary from tumor to tumor. Our inability to detect the submicroscopic changes which take place, for example, in the DNA during the preneoplastic or induction phase, hinders early detection. A neoplasm may require years to pass from the preneoplastic phase to the incipient phase, where it may then remain for months or years and be characterized by unclearly defined cytologic and histologic mutation. While in the incipient phase, neoplasms usually remain small, asymptomatic and seldom diagnostic, in the established phase most neoplasms are symptomatic, reach variable size, and can be diagnosed, though not without difficulty. Neoplasms may remain in the invasive phase for weeks or months prior to entering the metastatic phase. Most malignant soft tissue neoplasms are diagnosed in the invasive phase because they are symptomatic. In general, they are about 5 cm in size, exhibit characteristic growth patterns, and show identifiable cytologic abnormalities (1,5,6) (See Table 1).

The etiology of soft tissue tumors with a few exceptions (e.g., vinyl-chloride-induced hemangiosarcoma, genetically-transmitted nerve tumors, irradiation-induced sarcomas, and proliferative changes linked to trauma or lymphedema) is unknown and much work remains (1). The forms and types of soft tissue lesions are almost endless. Reliable assessment and identification of soft tissue tumors depend on the pathologist's ability to distinguish reactive, non-neoplastic lesions from benign and malignant neoplasms. Such an evaluation requires, among other things, knowledge of the names, definitions, and synonyms of various entities. Microscopic definition largely depends on the pathologist's visual memory. The pathologist should study the tissue pattern, the cell morphology, the appearance of the stroma and the products of the cells (see Table 1) as objectively as possible in conjunction with the age of the patient and the site and size of the tumor (1,2,7,8). Unfortunately there are very few criteria other than histologic pattern and cell morphology by which malignant soft tissue neoplasms can be diagnosed. Special stains, histochemical stains, electron microscopy, immunofluorescence microscopy and immunohistochemistry have not fulfilled every expectation. Although these techniques help to establish the histogenesis and the cell of origin of a given neoplasm, none can distinguish a malignant neoplasm from a benign one.

It is hoped that pathologists with full knowledge of the case will be able to examine generous and technically excellent tissue samples microscopically. In the case of malignant neoplasms a microscopic examination must end with assessment of the histologic grade (1,7). Once the sarcoma has been defined, the most important point is that the pathologist, surgeon, and oncologist agree on the best course of action and advise the patient accordingly (1,2,8).

REPORTED CASES OF SOFT TISSUE SARCOMAS

Nothing shows the clinical complexity and treacherous pathologic ambiguity of soft tissue tumors better than the reported cases of soft tissue sarcomas claimed to have been caused by herbicide exposure. I read the case reports from Sweden and the United States as they were published as letters to the editors, apparently without proper editorial scrutiny. In 1977, seven cases of soft tissue sarcomas were summarized as clinical observations in a Swedish publication (9); no information was disclosed as to the clinico-pathology of the cases and no pathologic data were included in the report. In 1979, a case-control study from Northern Sweden in the British Journal of Cancer (10) reported an approximately sixfold increased risk of soft tissue sarcoma (13 in a group of 52 patients). The cases were not described, not illustrated, and the type of sarcomas and how they were diagnosed were not included. In 1981, a case-referent study from Southern Sweden based on a cluster of patients with sarcoma was published in the British Journal of Industrial Medicine (11): it claimed a fivefold increased risk in patients exposed to herbicides. It is particularly surprising that more than 30% of the soft tissue sarcomas were diagnosed as leiomyosarcoma, a very rare form, and over 10% of the cases were listed as peripheral nerve sarcomas, an equally uncommon form of soft tissue sarcoma. Since the clinicopathology of the cases was poorly documented and the histopathology of the tumors was not illustrated, only the authors know whether the original seven cases reported in 1977 (9) were included in the 1979 and 1981 reports (10,11). One cannot escape the feeling that a number of cases were perhaps misclassified and that some tumors were perhaps not even sarcomas.

In January, 1981, the first three cases of soft tissue sarcoma in the United States were reported in a letter in Lancet (12). Two of these cases had been originally reported by different authors in separate case reports in the Journal of Occupational Medicine, (13,14). No detailed clinicopathological documentation was given in either the original case reports or in the Lancet letter. However, knowing the age at death (58, 53 and 49) of the patients, mode of presentation, and the histologic diagnoses, (malignant fibrous histiocytoma, fibrosarcoma, and liposarcoma) one would like to know whether non-sarcomatous neoplasms were ruled out.

The fourth American case was reported as malignant fibrous histiocytoma in the March, 1981 issue of Lancet (15). This letter also suggested that smokers who had substantial TCDD exposure may be at increased risk of fibrous soft tissue sarcoma.

The fifth American case was reported as a malignant peripheral nerve tumor in the June, 1981 issue of Lancet (16). Nothing was disclosed about the pathology of the case, and the authors of the letter failed to consider that the retroperitoneum is one site that very often poses difficult diagnostic challenges to pathologists (1).

The sixth and seventh American cases were published in the July, 1981 issue of Lancet (17). The letter presents a father and son combination and states that the father had "prolonged exposure" while the son "did not have a long latency period between exposure and diagnosis" (17). Again, there was no pathologic documentation of the cases. Furthermore, the credibility of the report is questionable, because there is no well-documented and reproducible study demonstrating that herbicides cause soft tissue sarcoma in animals. Without access to appropriate clinicopathologic material one can also equally speculate that tumors of these two patients, father and son, are genetically linked.

Finally, three cases of "thoracic soft tissue sarcoma in persons who served in Vietnam and were exposed to Agent Orange" were published as a letter in the May, 1982 issue of the New England Journal of Medicine (18). All three cases are from a Veterans Administration Hospital. Again, the reader has to take the authors', medical oncologists, words because none of the cases was documented pathologically.

It is perhaps appropriate to point out that the total Swedish experience with herbicide-induced soft tissue sarcomas amounts to about 16 cases (9,11). This occurred in a country with nearly 8,000,000 people who have an estimated annual incidence of 200 soft tissue sarcomas. Even more revealing is that in the United States with a population over 200,000,000, with an annual soft tissue sarcoma burden of 6,000, there are only 10 poorly-documented reported cases of soft tissue sarcoma that were linked to herbicides (12-18).

Without questioning whether the reported soft tissue sarcomas were actually sarcomas, I would like to do some crude non-statistical arithmetic. If there were 2,000,000 servicemen in and out of Vietnam, their expected incidence of soft tissue sarcoma would amount to 60 cases per year, by adhering to the standard incidence rate of soft tissue sarcoma in the American population. Yet, there were only three reported cases (18). It is also of interest that the studies from New Zealand (19), Europe (20), and Washington State (21), in which no convincing link was found between herbicides and soft tissue sarcoma, received little publicity.

No new case reports have appeared from Sweden since the original cases were reported in 1977, 1979 and again in 1981 (7-11). And we have yet to see a cause-and-effect case report from other northern countries, for example from Canada. Furthermore, I am not aware of any well-documented reports that can stand up to scientific scrutiny on increased number of sarcomas in Vietnam, where many of the 40,000,000 people were exposed to herbicides. Also, due to poor design, the results of certain experimental studies (22,23) cannot be taken at face value.

CONCLUSIONS

Being aware of the difficulties in documenting retrospective and anecdotal cases and assuming that the Swedish (9-11) and American (12-18) cases were all soft tissue sarcoma, I find the samples statistically insufficient for meaningful deduction. Recognizing the serious nature of linking herbicides with soft tissue sarcomas that may affect the lives of many people, I believe that --

1. we should urge the editors of medical and other professional journals to declare a moratorium on solitary case reports submitted as letters, and that
2. no manuscript should be published without the strictest editorial scrutiny or without complete clinicopathologic documentation for each case.

We can learn from past experiences and from reported cases but we need more well-designed experimental (24) and clinical (25) studies prior to reaching conclusions. As a beginning, I would like to see the creation of a national committee, a task force, composed of experts in clinical medicine, epidemiology, toxicology, chemistry, and pathology. This task force should --

1. be the final authority, after thorough scrutiny of all medical records and all pathological material, to decide whether any alleged case of cancer is linked to one or another type of herbicide;
2. uphold the highest professional and ethical standards; and
3. be assured under our medicolegal system that their conclusions will not lead to unwarranted litigations and unjustified malpractice claims.

REFERENCES

1. Hajdu, S.I. Pathology of Soft Tissue Tumors. Lea & Febiger, Philadelphia, 1979.

2. Hajdu, S.I. Differential Diagnosis of Soft Tissue and Bone Tumors. Lea & Febiger, Philadelphia, 1984.

3. Hajdu, S.I. Soft tissue sarcomas. Classification and natural history. CA-A Cancer Journal for Clinicians 31:271-280, 1981.

4. Hajdu, S.I., Lemos, L.B., Kozakewich, H., Helson, L. and Beattie, E.J. Growth pattern and differentiation of human soft tissue sarcomas in nude mice. Cancer 47:90-98, 1981.

5. Hajdu, S.I. and Hajdu, E.O. Cytopathology of Sarcomas and Other Non-epithelial Malignant Tumors. W.B. Saunders, Philadelphia, 1976.

6. Hajdu, S.I., Bean, M.A., Fogh, J., Hajdu, E.O. and Ricci, A. Papanicolaou smear of cultured human tumor cells. Acta Cytologica 18: 327-332, 1974.

7. Russel, W.O., Cohen, J., Enzinger, F., Hajdu, S.I. et al. A clinical and pathological staging system for soft tissue sarcomas. Cancer 40:1561-1570, 1977.

8. Fine, G., Hajdu, S.I., Morton, D.L., Eilber, F.R., Suit, H.D. and Weiss, S.W. Soft Tissue Sarcomas. Classification and Treatment (A symposium). S.C. Sommers and P.P. Rosen (editors). Pathology Annual 17:155-195, 1982.

9. Hardell, L. Soft tissue sarcomas and exposure to phenoxyacetic acids - a clinical observation. Lakartidningen 74:2753-2754, 1977.

10. Hardell, L., Sandstrom, A. Case-control study: soft-tissue sarcomas and exposure to phenoxyacetic acids or chlorophenols. British Journal of Cancer 39:711-717, 1979.

11. Eriksson, M., Hardell, L., Berg, N.O., Moller, T. and Axelson, O. Soft tissue sarcomas and exposure to chemical substances: a case-referent study. British Journal of Industrial Medicine 38: 27-33, 1981.

12. Honchar, P.A. and Halperin, W.E. 2,4,5-T, trichlorophenol, and soft tissue sarcoma (Letter). Lancet i:268-269, 1981.

13. Zack, J.A. and Suskind, R.R. The mortality experience of workers exposed to tetrachlorodibenzodioxin in a trichlorophenol process accident. Journal of Occupational Medicine 22:11-14, 1980.

14. Cook, R.R., Townsend, J.C., Ott, G. and Silverstein, I.G. Mortality experience of employees exposed to 2,3,7,8-tetra-chlorodibenzo-p-dioxin (TCDD). Journal of Occupational Medicine 22:530-532, 1980.

15. Cook, R. Dioxin, chloracne, and soft tissue sarcoma (Letter). Lancet i:618-619, 1981.

16. Moses, M. and Selikoff, I.J. Soft tissue sarcomas, phenoxy herbicides, and chlorinated phenols (Letter). Lancet i:1370, 1981.

17. Johnson, F.E., Kugler, M.A. and Brown, S.M. Soft tissue sarcomas and chlorinated phenols (Letter). Lancet ii:40, 1981.

18. Sarma, P.R. and Jacobs, J. Thoracic soft-tissue sarcoma in Vietnam veterans exposed to agent orange (Letter). New England Journal of Medicine 306:1109, 1982.

19. Smith, A.H., Matheson, D.P., Fisher, D.O. and Chapman, C.J. Preliminary report of reproductive outcomes among pesticide applicators using 2,4,5-T. New Zealand Medical Journal 680:177-179, 1981.

20. Riihimaki, V., Asp, S. and Hernberg, S. Mortality of 2,4-dichlorophenoxyacetic acid and 2,4,5-trichlorophenoxyacetic acid herbicide applicators in Finland. Scandinavian Journal of Work Environment and Health 8:37-42, 1982.

21. Milham, S. Herbicides, occupation, and cancer (Letter). Lancet i:1464-1465, 1982.

22. Van Miller, J.P., Lalich, J.J. and Allen, J.R. Increased incidence of neoplasms in rats exposed to low levels of 2,3,7,8-tetrachlorodibenzo-p-dioxin. Chemosphere 9:537-544, 1977.

23. Kociba, R.J., Keyes, D.G., Beyer, J.E., Carreon, R.M. et al. Results of a two-year chronic toxicity and oncogenicity study of 2,3,7,8-tetrachlorodibenzo-p-dioxin in rats. Toxicology and Applied Pharmacology 46:279-303, 1978.

24. Bioassay of 2,3,7,8-tetrachlorodibenzo-p-dioxin for possible carcinogenicity. National Institutes of Health. DHHS Publication No. (NIH) 80-1757.

25. Berry, D.L., Slaga, T.J., DiGiovanni, J. and Juchau, M.R. Studies with chlorinated dibenzo-p-dioxins, polybrominated biphenyls, and polychlorinated biphenyls in a two-stage system of mouse skin tumorigenesis: potent anticarcinogenic effects. Annals of the New York Academy of Sciences 77:405-414, 1979.

REVIEW OF EXPOSURE AND PATHOLOGY DATA FOR SEVEN CASES REPORTED AS SOFT TISSUE SARCOMA AMONG PERSONS OCCUPATIONALLY EXPOSED TO DIOXIN-CONTAMINATED HERBICIDES

Marilyn, A. Fingerhut, Ph.D*
William E. Halperin, M.D., M.P.H.*
Patricia A. Honchar, Ph.D*
Alexander B. Smith, M.D., M.S.*
David H. Groth, M.D.*
William O. Russell, M.D.**

*National Institute for Occupational Safety and Health
Cincinnati, Ohio 45226

**North Ridge General Hospital and Cancer Foundation,
Fort Lauderdale, Florida 33334

From Public Health Risks of the Dioxins, proceedings of a symposium held on October 19-20, 1983 at The Rockefeller University, New York City. Edited by William W. Lowrance. Published by William Kaufmann, Los Altos, California, 1984.

Data from two Swedish studies (1,2) and a review of four U.S. studies (3) of dioxin-exposed workers have suggested that occupational exposure to dioxin-contaminated products is associated with an increased risk of soft tissue sarcoma (STS). Soft tissue sarcomas are malignant neoplasms arising throughout the body from mesenchymal supporting tissue other than bone (4). The principal types of histopathologic subtypes and their frequencies are listed in Table 1.

Table 1. Histopathologic varieties of sarcoma of soft tissue.*

Histopathological Variety	Percentage of Total Cases
Rhabdomyosarcoma	19.2
Fibrosarcoma	19.0
Liposarcoma	18.2
Malignant fibrohistiocytoma	10.5
Sarcoma of Soft Tissue, type unspecified	10.0
Synovial sarcoma	6.9
Leiomyosarcoma	6.5
Malignant schwannoma	4.9
Angiosarcoma	2.7
Other types	1.9

* Table presented in Suit, H.D., Sarcoma of soft tissue, CA-A Cancer Journal for Clinicians 1978; 28:284-295.

The percentages were derived from a study of 1,215 sarcomas by the Task Force for Sarcoma of Soft Tissue of the American Joint Committee on Cancer (5).

Dioxins are generated as unintended contaminants during the chemical manufacture of chlorinated phenoxy acetic acids and chlorinated phenols. The most toxic of the 75 dioxin isomers, 2,3,7,8-tetrachlorodibenzodioxin (2,3,7,8-TCDD), is generated during the production of trichlorophenol (TCP) and its derivative, the herbicide 2,4,5-trichlorophenoxyacetic acid (2,4,5-T). Industrial workers who produce chlorinated phenoxy acetic acids and chlorophenols, and herbicide applicators who spray these products, potentially have been exposed to the products and their dioxin contaminants. In 1977, several cases of soft tissue sarcoma were reported among Swedish lumberjacks who had had prior exposure to phenoxy herbicides. This clinical observation led researchers in Sweden to conduct two separate case control studies (1,2). Both studies found that persons with occupational exposure to phenoxy acids or chlorophenols had a fivefold increased risk of developing a soft tissue sarcoma.

At about the same time, four studies (6-9) were conducted in U.S. manufacturing plants of workers exposed to the herbicide 2,4,5-T and its chemical precursor, trichlorophenol. The dioxin isomer contaminating these products is the most toxic form, 2,3,7,8-TCDD. None of the four studies found any statistically significant excess in total mortality or in death from cancer which was attributed to dioxin exposure. However, each cohort was small and of insufficient statistical power to allow adequate evaluation of rare causes of death. Honchar (3) reviewed the deaths in the four cohorts, and found that three (2.9%) of the total 105 deaths in the merged cohorts were reported to be from soft tissue sarcoma. Based on national statistics, only 0.07% of deaths was expected to be due to this cause. Table 2 lists these three deaths as Cases 1 to 3. Subsequently, Dr. Ralph Cook of Dow Chemical Company reported a fourth, living person in one of these cohorts as having a soft tissue sarcoma (10). This individual, Case 4, is now deceased. Dr. Marion Moses of Mt. Sinai reported an additional case of soft tissue sarcoma in an individual (Case 5) employed at one of the same four chemical manufacturing sites (11). Subsequently, two additional persons, Cases 6 and 7, were reported to have soft tissue sarcomas (12). They worked at a chemical manufacturing site which produced 2,4,5-T, but which had not been previously studied, and their exposures had not been confirmed at the time of the reports.

Table 2. Exposure data from original publications for the seven U.S. workers.

Case Number	Type of Exposure	Type of Report	Reference
1	TCP	Mortality Study	Zack, J. and Suskind, R.S. *Journal of Occupational Medicine* 1980; 22:11-14.
2	2,4,5-T	Mortality Study	Zack, J. and Gaffey, W.R. *Environmental Science Research* 1983; 26:575-591.
3	TCP	Mortality Study	Cook, R.R. et al. *Journal of Occupational Medicine* 1980; 22:530-32.
4	TCP	Case Report	Cook, R.R. *Lancet* 1981; i:618-619.
5	TCP, 2,4,5-T	Case Report	Moses, M. and Selikoff, I.J. *Lancet* 1981; i: 1370.
6	Chlorophenols	Case Report	Johnson, F.E. et al. *Lancet* 1981; i:40.
7	Chlorophenols	Case Report	Johnson, F.E. et al. *Lancet* 1981; i:40.

We have obtained detailed employment records, medical and pathological reports, tissue specimens and death certificates for these seven individuals and present here detailed descriptions of our occupational and pathologic reviews. We hope, through this report, to focus attention on the problems associated with the study of soft tissue sarcomas in dioxin-exposed populations. In addition to our work at the National Institute for Occupational Safety and Health (NIOSH), studies of the health effects of dioxin are underway at the Centers for Disease Control (CDC), the Veterans Administration (VA), the National Institutes of Health (NIH), and various state agencies in this country. Other organizations, both here and abroad, are also conducting studies of soft tissue sarcoma, and the International Agency for Research on Cancer (IARC) of the World Health Organization is considering an International Dioxin Registry.

EXPOSURE DATA FOR THE SEVEN CASES

Table 3 presents the exposure history of each individual. We used a rigorous criterion of exposure. An individual was considered exposed to products contaminated with 2,3,7,8-TCDD if he had a company record of assignment to a department producing trichlorophenol or 2,4,5-T.

Table 3. Work history information for the seven U.S. workers.

Case Number	Facility	Years of Employment	Job Title	Duration of Exposure (Years)
1	A	1946-1978	TCP Operator	1.9
2	A	1946-1972	2,4,5-T Operator	2.0
3	B	1950-1975	Maintenance	3.5
4	B	1951-1982	TCP Operator Plant Mechanic (TCP Department)	19.0
5	A	1943-1975	Maintenance Service	--
6	C	1978-1980	Production Worker	--
7	C	1951-1980	Production Worker	11 days

The work histories confirmed the original reports that Cases 1, 2, and 4 were production workers assigned to the trichlorophenol or 2,4,5-T departments. Case 3 was a maintenance worker identified by the company as assigned to a building where trichlorophenol was produced. Case 5 was a maintenance and service worker for 32 years in a chemical manufacturing site which produced trichlorophenol, 2,4,5-T, and many other chemicals, but he had no record of specific assignment to a trichlorophenol or 2,4,5-T department. Case 6 worked two and one-half years in a plant which made 2,4,5-T. He was a production worker but had no record of assignment to the 2,4,5-T area. Case 7 is the father of Case 6 and was a production worker at the same facility for 29 years. His work history showed no record of assignment to the 2,4,5-T area, although he did work for 11 days following hire in 1951 in a pentachlorophenol area. Pentachlorophenol is contaminated with hexa-, hepta- and octachlorinated isomers of dioxin, but does not contain 2,3,7,8-TCDD. These last three individuals illustrate some of the difficulties encountered in attempting to verify exposure in chemical workers. In general, production workers do have records of regular assignment to specific departments within complex chemical manufacturing plants, although their records do not show temporary changes in assignments. However, maintenance workers such as Case 5 often work throughout an entire facility and never have recorded assignments in specific departments. From our examination of their work histories we concluded that these three workers did not meet our stringent criterion of exposure, since they had no records of assignment to a 2,4,5-T or trichlorophenol process. However, absence of such documentation does not eliminate the possibility that they were exposed to dioxin contaminated products during their routine activities.

PATHOLOGY DATA FOR THE SEVEN CASES

Table 4 summarizes clinical observations obtained from company medical records, hospital records, and pathology reports for the seven individuals. All cases were reported to have soft tissue sarcomas.

Table 4. Medical information for the seven U.S. workers.

Case Number	Tumor	Onset of Tumor	Age at Onset	Chloracne
1	Malignant fibrous histiocytoma	1978	58	yes
2	Adenocarcinoma, bladder	1958	34	yes
	Liposarcoma, rectum	1972	49	
3	Fibrosarcoma	1973	52	dermatitis
4	Malignant fibrous histiocytoma	1979	58	yes
5	Neurogenic sarcoma	1980	57	no

Table 4. Continued.

Case Number	Tumor	Onset of Tumor	Age at Onset	Chloracne
6	Fibrosarcomatous mesothelioma	1980	33	no
7	Liposarcoma	1980	53	no

Cases 1, 2, and 4 had records of diagnosed chloracne. Case 3 had dermatitis of the face and neck while assigned to the trichlorophenol building during a period when 49 trichlorophenol workers developed chloracne (8). But no chloracne was found in the records of Cases 5, 6, and 7.

Table 5 presents data on the latency of the tumors (time from first exposure to diagnosis of the tumor) for each individual.

Table 5. Latency of tumors for seven U.S. workers.

Case Number	First Assignment to 2,4,5-T or TCP Department	STS Tumor Onset	Latency* (Years)	Age at Onset
1	1949	1978	29	58
2	1950	1971	21	49
3	1964	1973	9	52
4	1951	1979	28	58
5	never	1980	N.A.**	57
6	never	1980	N.A.	33
7	never	1980	N.A.	53

* Time from first exposure to onset of tumor.
** Not applicable.

Adequate latency for development of cancer was experienced by the four individuals with records of assignment to trichlorophenol or 2,4,5-T departments (Cases 1 to 4). Since no records of specific departmental assignment to TCP or 2,4,5-T departments exist for the remaining three individuals, latency cannot be assigned for exposure to 2,3,7,8-TCDD.

DEATH CERTIFICATE DATA VS. PATHOLOGIC DIAGNOSES AND ICD CODING CATEGORIES

Table 6 compares the diagnoses made by the attending pathologists with the information on the death certificates.

Table 6. Comparison of death certificate data with original pathologic diagnoses for the seven U.S. workers.

Case Number	Death Certificate	Original Pathology
1	Malignant fibrous histiocytoma	Malignant fibrous histiocytoma
2	Liposarcoma	Liposarcoma
3	Fibrosarcoma	Fibrosarcoma
4	Malignant fibrous histiocytoma	Malignant fibrous histiocytoma
5	Carcinomatosis	Myxoid neurogenic sarcoma
6	Metastatic mesothelioma	Fibrosarcoma consistent with fibrosarcomatous mesothelioma
7	N.A.*	Myxoid liposarcoma

* Not applicable. Individual is alive.

Of the six deceased cases, all of whom had hospital diagnoses of soft tissue sarcoma, only four had soft tissue sarcoma noted on their death certificates (Cases 1 to 4). A similar finding was noted in a study of the accuracy of death certificates conducted by Percy et al. in 1981 (13). They reviewed almost 50,000 hospital diagnoses and death certificates and found that of 252 soft tissue sarcomas diagnosed in hospital records, only 142 death certificates for those individuals (55%) reported the soft tissue sarcomas. Hence using death certificates as the primary source of cases may lead to underascertainment of cases.

We found in our review that all four death certificates with notations of soft tissue sarcoma (Cases 1 to 4) belonged to persons who had diagnoses of soft tissue sarcoma made by the attending pathologists. In contrast, Percy (13) found in her large study that only 56% of death certificates with notation of soft tissue sarcoma were supported by hospital records with the diagnoses of soft tissue sarcoma. Hence using death certificates as the primary source of cases and not confirming hospital or pathologic information may lead to overascertainment. Researchers conducting epidemiologic studies should be fully aware of possible underascertainment or overascertainment.

Table 7 illustrates another issue in the use of death certificate data: consistency in classification by nosologists, the experts who code cause of death from death certificates according to the rules of the International Classification of Disease System (14).

Table 7. Comparison of death certificate information and nosologic coding of the information.

Case Number	Death Certificate	ICD Category* Original Publication**	ICD Category* Honchar Review***
1	Malignant fibrous histiocytoma	173.9	171

* World Health Organization, Manual of the International Statistical Classification Diseases, Injuries and Causes of Death, Geneva. Ninth Revision, 1975.

** Zack, J.A. and Suskind, R.S. Journal of Occupational Medicine 1980; 22:11-14.

*** Honchar, P.A. and Halperin, W.E. Lancet 1981; i:268-269.

Case 1 was reported in the original publication as having a malignant fibrous histiocytoma, which was coded in ICD Category 173.9, "Other malignant neoplasm of skin, site unspecified". Honchar indicated in her review (3) that the choice of her nosologist was ICD 171, "Malignant neoplasm of connective and other soft tissue". Misclassification in ICD categories could introduce significant error. The data in Table 7 suggest that coding by more than one trained nosologist may be desirable.

Another important issue of classification not addressed by this review has serious implications for studies of soft tissue sarcoma. The International Classification of Disease System is a site-oriented classification scheme. Therefore, sarcomas that develop in the parenchymatous organs such as the uterus or stomach are coded into the ICD categories for the organ site. Only soft tissue sarcomas of the supporting tissue of the body not specified as arising in organs are coded in the ICD Category 171. It is possible, therefore, to select soft tissue sarcoma cases for case control studies in two ways. For the early Swedish studies (1,2) cases were selected by their histopathologic characteristics; approximately 60% of these cases are coded in categories other than ICD 171 (15). A current case control study in New Zealand (16) includes only soft tissue sarcoma cases coded as ICD 171. Such differences in study design must be recognized when evaluating results. The problem presented by the ICD dual classification scheme is slightly different for cohort mortality studies. It is possible to compare sarcomas coded as ICD 171 with those arising in the national population; however, we are not aware of population rates for soft tissue sarcoma coded into other ICD categories.

ORIGINAL REPORT DATA VS. NEW REVIEW DATA

Table 8 compares the original pathologic diagnoses for the seven individuals with the diagnoses selected in two independent pathologic reviews of tissue specimens.

Table 8. Comparison of the original pathology reports with reports of two reviewers.*

Case Number	Original Pathology	Review #1	Review #2
1	Malignant fibrous histiocytoma	Malignant fibrous histiocytoma	Malignant fibrous histiocytoma
2	Invasive pleomorphic liposarcoma	Poorly differentiated carcinoma	Carcinoma, poorly differentiated
3	Fibrosarcoma	Clear cell carcinoma with spindling, renal	Spindle cell renal carcinoma
4	Malignant fibrous histiocytoma	Malignant fibrous histiocytoma	Malignant schwannoma
5	Myxoid neurogenic sarcoma	Leiomyosarcoma	Malignant fibro-histiocytoma
6	Fibrosarcoma	Malignant schwannoma	Malignant schwannoma
7	Liposarcoma	Myxoid liposarcoma	Myxoid liposarcoma

* The tissue specimens were reviewed by one of the authors (W.O.R.) and also by Franz M. Enzinger, M.D. and Sharon M. Weiss, M.D. of the Armed Forces Institute of Pathology, Washington, DC.

Although all seven individuals received diagnoses of soft tissue sarcoma by the original pathologists, only five of the cases were diagnosed as soft tissue sarcoma in each of the two reviews. In both reviews Cases 2 and 3 were identified as carcinomas rather than sarcomas. These results indicate that review of tissue specimens by pathologists with expertise in diagnosing soft tissue sarcomas is necessary for epidemiologic studies. A difficult problem arises for cohort mortality studies because there is no comparison group for cases ascertained by expert pathologists. Rates of disease commonly used for comparison are those taken from compilations of population rates based on information from death certificates.

The adequacy of diagnosis of histological subtypes of soft tissue sarcoma is another issue raised by the data in Table 8. Although the original pathologists and the reviewers all agreed that five individuals had soft tissue sarcomas, they agreed on the histological subtype for only two individuals, Cases 1 and 7. Even the expert reviewers disagreed on subtype diagnoses for Cases 4 and 5. Because of the difficulties involved in diagnosing the histological subtypes, particularly when only a limited amount of tissue is available for review, we recommend that epidemiologic studies continue to focus on the outcome soft tissue sarcoma, and assess the distribution of subtypes only cautiously and with a recognition of the limitations involved.

The next two tables summarize the exposure histories and the pathologic findings and compare them to the information published in the original reports. Table 9 presents data for the four production workers identified in company studies (6-10) as 2,4,5-T or trichlorophenol-exposed workers.

Table 9. Comparison of original reports with results from this review for production workers identified by companies as TCP* or 2,4,5-T workers.

	Reported Exposure		Cause of death	
Case Number	Original Report	This Review	Original Report	This Review
1	TCP	TCP	STS	STS
2	2,4,5-T	2,4,5-T	STS	Carcinoma
3	TCP	TCP	STS	Carcinoma
4	TCP	TCP	STS	STS

* Terms: TCP, trichlorophenol; 2,4,5-T, 2,4,5-trichlorophenoxyacetic acid; STS, soft tissue sarcoma.

The reported exposures were confirmed by documentation that the four individuals had been assigned to 2,4,5-T or TCP processes. All had been reported as having soft tissue sarcomas, but we report that the two expert pathologists confirm that only two individuals (Cases 1 and 4) had soft tissue sarcomas. Both of these individuals worked in trichlorophenol processes, had diagnosed chloracne, and were present during incidents of unusual operating conditions with probable exposure to 2,3,7,8-TCDD. The histological subtype of malignant fibrous histiocytoma was selected by all three pathologists for Case 1. Case 4 also received a diagnosis of malignant fibrous histiocytoma from the original pathologist and one reviewer. The second reviewer identified the subtype as a neurogenic schwannoma.

Table 10 presents the same comparison for the three individuals originally described in case reports by physicians (11,12).

Table 10. Comparison of original reports with results from this review for production workers originally described as case reports by physicians.

Case Number	Reported Exposure: Original Report	Reported Exposure: This Review	Pathologic Diagnosis: Original Report	Pathologic Diagnosis: This Review
5	Possibly 2,4,5-T,* TCP	None	STS	STS
6	Possibly chlorophenols	None	STS	STS
7	Possibly chlorophenols	PCP	STS	STS

* Terms: 2,4,5-T, 2,4,5-trichlorophenoxyacetic acid; TCP, trichlorophenol; PCP, pentachlorophenol; STS, soft tissue sarcoma.

Review of the tissue specimens for all three workers were confirmed that they had soft tissue sarcomas. Examination of the work history records of the individuals did not document assignment to 2,4,5-T or trichlorophenol departments. However, Case 7 worked briefly upon hire in a pentachlorophenol department.

CONCLUSION

Swedish research (1,2) has drawn attention to the issue of soft tissue sarcoma and dioxin exposure. While each of the original four U.S. studies (6-9) reported non-positive findings, a reanalysis (3) and further reports (10,11) strengthened the suspicion that an association might exist. Honchar (3) and Cook (10) called for detailed examination of the data. We present the detailed assessment of exposure history and pathologic diagnosis and report that of the four soft tissue sarcoma cases identified by death certificates among the TCP and 2,4,5-T workers, two are confirmed following pathologic review of tissues as soft tissue sarcoma. It is not possible to use our data to draw a definitive conclusion regarding the suspected association because we are not aware of any population rates for soft tissue sarcoma based upon pathologic review of tissue specimens.

Cases 5, 6 and 7 were originally described in the literature as individual case reports. All three are confirmed as cases of soft tissue sarcoma. Our review of their work histories did not find any record of assignment to 2,4,5-T or trichlorophenol departments, although Case 7 worked briefly in a pentachlorophenol department, a product contaminated with isomers of dioxin considered to be less toxic than 2,3,7,8-TCDD. Although these workers do not meet our stringent criterion of exposure by virtue of assignment to a 2,4,5-T or TCP department, we cannot exclude the possibility that they had undocumented contact with 2,4,5-T or TCP.

This review examines the complexities in evaluating the dioxin exposure of individuals and the pathologic diagnoses of soft tissue sarcoma. It emphasizes the need for carefully-designed large epidemiologic studies to adequately assess whether there is an association of soft tissue sarcoma with exposure to dioxin-contaminated products. We have shown that the use of the death certificate, even with the original pathologic diagnosis may lead to overascertainment or underascertainment of cases of soft tissue sarcoma. Finally, we reassert that further research, particularly the completion of National Institute for Occupational Safety and Health studies based upon approximately 6,000 U.S. workers, is necessary to confirm or refute the association first suggested in the Swedish studies.

ABSTRACT

We have reviewed medical and exposure records and pathology specimens of seven U.S. chemical workers reported in the literature to be cases of soft tissue sarcoma (STS) and to have had dioxin exposure. The cases were of interest because two Swedish studies demonstrated a strong association between STS and dioxin exposure.

Four U.S. workers from four small mortality cohorts had been reported to have died of soft tissue sarcoma. We found that these individuals had employment records of assignment to production of TCP and 2,4,5-T which are contaminated with the most toxic dioxin isomer, 2,3,7,8-tetrachlorodibenzo-p-dioxin (2,3,7,8-TCDD), but pathologic review of their tissue indicates that only two are cases of soft tissue sarcoma. Three additional individuals had been described by physicians as case reports. They are confirmed as cases of soft tissue sarcoma, but we did not find any record of assignment to these production departments. One person worked briefly in the production of pentachlorophenol, which is contaminated with other isomers of dioxin. We suggest that identification of cases of STS through either death certificates alone or through pathology records can lead to errors of ascertainment. We reassert that further research, particularly the completion of studies by the National Institute for Occupational Safety and Health based upon approximately 6,000 U.S. workers, is necessary to confirm or refute the association first suggested in the Swedish studies.

REFERENCES

1. Hardell, L. and Sandstrom, A. Case-Control Study: Soft-Tissue Sarcomas and Exposure to Phenoxyacetic Acids or Chlorophenols. British Journal of Cancer 1979; 39:711-717.

2. Eriksson, M., Berg, N., Hardell, L., Moller, T. and Alexson, O. Soft-Tissue Sarcomas and Exposure to Chemical Substances: A Case-Referent Study. British Journal of Industrial Medicine 1981; 38:27-33.

3. Honchar, P.A., and Halperin, W.E. 2,4,5-Trichlorophenol and Soft Tissue Sarcoma. Lancet 1981; i:268-269.

4. Suit, H.D. Sarcoma of Soft Tissue. CA-A Cancer Journal for Clinicians 1978; 28:284-295.

5. Russell, W.O. et al. A Clinical and Pathological Staging System for Soft Tissue Sarcomas. Cancer 1977; 40:1562-1570.

6. Zack, J.A. and Suskind, R.S. The Mortality Experience of Workers Exposed to Tetrachlorodibenzodioxin in a Trichlorophenol Process. Journal of Occupational Medicine 1980; 22:11-14.

7. Ott, M.G., Holder, B.B. and Olson, R.D. A Mortality Analysis of Employees Engaged in the Manufacture of 2,4,5-Trichlorophenoxyacetic Acid. Journal of Occupational Medicine 1980; 22:47-50.

8. Cook, R.R., Townsend, J.C. and Ott, M.G. Mortality Experience of Employees Exposed to 2,3,7,8-Tetrachlorodibenzo-p-dioxin (TCDD). Journal of Occupational Medicine 1980; 22:530-532.

9. Zack, J.A. and Gaffey, W.R. A Mortality Study of Workers Employed at The Monsanto Company Plant in Nitro, West Virginia. Environmental Science Research 1983; 26:575-591.

10. Cook, R.R. Dioxin, Chloracne and Soft Tissue Sarcoma. Lancet 1981; i:618-619.

11. Moses, M. and Selikoff, I.J. Soft Tissue Sarcomas, Phenoxy Herbicides and Chlorinated Phenols. Lancet 1981; i:1370.

12. Johnson, F.E., Kugler, N.A. and Brown, S.M. Soft Tissue Sarcomas and Chlorinated Phenols. Lancet 1981; i:40.

13. Percy, C., Stanek, E. and Gloeckler, L. Accuracy of Cancer Death Certificates and Its Effect on Cancer Mortality Statistics. American Journal of Public Health 1981; 71:242-250.

14. World Health Organization, Manual of the International Statistical Classification of Diseases, Injuries and Causes of Death (1975) Ninth Revision. Geneva, 1977.

15. Summarized from Exhibit 1299, prepared by L. Hardell for the U.S. Environmental Protection Agency, October 24, 1979.

16. Smith, A.H. et al. Do Agricultural Chemicals Cause Soft Tissue Sarcoma? Initial Findings of a Case Control Study in New Zealand. Community Health Studies 1982; 6:114-119.

SOFT TISSUE SARCOMA AND DIOXINS: PUTTING THE DATA INTO PERSPECTIVE

Ralph R. Cook, M.D., M.P.H.
Janice B. Cartmill, R.N.

Department of Epidemiology
Dow Chemical U.S.A.
Midland, Michigan 48640

From *Public Health Risks of the Dioxins*, proceedings of a symposium held on October 19-20, 1983 at The Rockefeller University, New York City. Edited by William W. Lowrance. Copyright, The Rockefeller University, 1984. Published by William Kaufmann, Los Altos, California.

There has been a lot of confusion, controversy, and fear about the possible human health effects due to dioxins. Part of this confusion is due to the multiplicity of different chemicals that have been placed under the general heading "dioxin." Part is due to an inability to document precisely who among the general population, Vietnam veterans, and even select employee groups were or were not actually exposed, and to what, when, how much, and for how long. And part is due to the variability, and often the vagueness, of the health claims. During this dioxin controversy, everything from schizophrenia, to cancer, to birth defects has been attributed to dioxin exposure.

Most persistent has been the claim that 2,3,7,8-TCDD causes soft tissue sarcomas. This claim is based on three sources of information: the case-control studies in Sweden by Dr. Hardell and associates (1,2), the seven case reports among U.S. employees of Dow and Monsanto (3-6), and the excess connective tissue cancer reported among white women in Midland County, Michigan, site of a Dow Chemical manufacturing facility (7). Based on the above information, some have concluded that dioxins cause soft tissue sarcomas.

I disagree. I think the topic needs to be studied further, because I believe there are too many inconsistencies and too many potentially false assumptions to support a cause-effect relationship between dioxins and soft tissue sarcomas. Let me discuss each of these three sources in turn.

SWEDISH CASE-CONTROL STUDIES

These studies have been severely criticized by academic, government, and industry scientists both here and abroad (8-10). Because of the way the data were collected, risk estimates may have been artificially elevated. The reports themselves present data that support the possibility of both recall and observer bias. More importantly, no one else has been able to replicate Dr. Hardell's findings (11-13).

Of interest are the recent reports by Eklund, 1983 from the

Karolinska Institute which indicate that Swedish farmers and other users of phenoxy herbicides have not been at increased risk of soft tissue tumors. This is also true for Hodgkin's and non-Hodgkin's lymphoma, two other diseases which Hardell reported to be associated with phenoxy herbicide and chlorophenols exposure (14-15). Quite the contrary, the farmers have experienced slightly fewer cases of tumors or lymphomas than expected when compared with the total Swedish population (Table 1).

Table 1. Observed and expected cases of soft tissue tumors, Hodgkin's, and non-Hodgkin's lymphomas among male Swedish farmers and gardeners, 1961-1979.

	Soft Tissue Tumors	Hodgkin's Lymphoma	Non-Hodgkin's Lymphoma	Total
Observed	221	274	743	1238
Expected	226.2	290.8	767.4	1284.4
Ratio O/E	0.98	0.94	0.97	0.96

Adapted from: G. Eklund, (15).

Since about 15% of Swedish farmers used phenoxy herbicides, the only way the Hardell studies and the Karolinska Institute research can both be correct is if Swedish farmers not exposed to phenoxy herbicides develop these malignancies at somewhere between one-quarter and one-half the rate of the entire Swedish population. That seems unlikely. Until these and other inconsistencies are resolved, I think Dr. Hardell's work is, at best, intriguing. It does not allow us to make conclusions about cause and effect.

Before I discuss the case reports and the statistics from Midland County, I would like to make some comments about soft tissue sarcomas and connective tissue cancers.

SOFT TISSUE SARCOMA VS. MALIGNANT NEOPLASM OF CONNECTIVE AND OTHER SOFT TISSUE

Many think that these terms are synonyms. They are not. They overlap, but they are not interchangeable. They are not equivalent. Soft Tissue Sarcoma includes a group of tumors defined by their histopathology, by how they look under the microscope. Malignant Neoplasm of Connective and Other Soft Tissue is a category in the International Classification of Diseases.

The International Classification of Diseases (ICD), utilized in producing national mortality tables, was developed for statistical reporting purposes (16). With this objective in mind, it was desirable to construct a limited number of categories that encompassed the entire range of morbid conditions. As a consequence, rare conditions tend to be grouped together in less homogeneous categories that may be inconsistent from a histologic, anatomic, or etiologic viewpoint.

The ICD category Malignant Neoplasm of Connective and Other Soft Tissue includes a group of diagnostic entities that is neither limited to all cancers sharing a common histology (at least six major tissue types of mesodermal origin are included) nor represents cancers of a common anatomical site. Rather, histologically-defined sarcomas are included when the anatomical site is the extremities or is not specified. Histologically-defined sarcomas of specific parenchymal organs are coded to cancer of that site (17).

For example, a leiomyosarcoma of the uterus (given this as the underlying cause of death) could appear on the death certificate three different ways: leiomyosarcoma, leiomyosarcoma of the uterus, or cancer of the uterus (Table 2). This same histological entity would be handled differently from a nosological perspective depending on what the physician wrote on the death certificate and where he wrote it. If the entry in Part I were cancer of the uterus, it would fall into ICD-8 Code 182, Other Malignant Neoplasm of the Uterus. If the entry were leiomyosarcoma of the uterus, the ICD-8 Code would also be 182. Only if the entry were simply leiomyosarcoma, site unspecified, would the ICD-8 Code be 171, Malignant Neoplasm of

Connective and Other Soft Tissue. Furthermore, if the physician had written any one of these three entries into Part II of the death certificate, probably none of the above codes would be used for underlying cause of death.

Table 2. Effect of a physician's death certificate entry on ICD coding.

Histopathology: Leiomyosarcoma of the Uterus

Diagnosis on Death Certificate	ICD-8 Code	ICD-8 Category
Cancer of the uterus	182	Other Malignant Neoplasm of Uterus
Leiomyosarcoma of the uterus	182	Other Malignant Neoplasm of Uterus
Leiomyosarcoma	171	Malignant Neoplasm of Connective and Other Soft Tissue

Based on the Third National Cancer Survey data, 35 of 186 leiomyosarcomas among white men, and 49 of 437 leiomyosarcomas among white women would be coded as Malignant Neoplasm of Connective and Other Soft Tissue (18). It is obvious that one cannot calculate relative risks using a histopathological classification for the exposed and an ICD classification for the controls. Such a procedure automatically builds a large bias into any risk calculations producing a gross overestimate of relative risk.

Sound confusing? Let me further complicate the picture. The rules of ICD coding differ depending on which revision is in place at the time of death (Table 3).

Table 3. International Classification of Diseases (ICD) revision codes and years of use.

Revision Number	Years Used
ICD-7	1958 - 1967
ICD-8	1968 - 1978
ICD-9	1979 - present

ICD-7 covered the years 1958 through 1967; ICD-8, 1968 through 1978; and ICD-9 went into effect in 1979 (19-20). Not only can the statistics for rare forms of cancers be a problem within a given ICD revision, but great caution must be exercised when trying to interpret data between different ICD revisions.

For example, as shown in Table 4, a hemangiopericytoma would be classified as a 228, a benign neoplasm under ICD-7 rules; as a 171.9 Malignant Neoplasm of Connective and Other Soft Tissue under ICD-8; and as a 238.1, Neoplasm of Uncertain Behavior, under ICD-9.

Table 4. Effect of ICD revisions on coding of hemangiopericytoma.

Diagnosis: Hemangiopericytoma

ICD Revision	ICD Code	ICD Category
ICD-7	228	Hemangioma and Lymphangioma (This is a benign neoplasm category.)
ICD-8	171.9	Malignant Neoplasm of Connective and Other Soft Tissue, Site Unspecified
ICD-9	238.1	Neoplasm of Uncertain Behavior of Other and Unspecified Sites and Tissues - Connective and Other Soft Tissue

Another example will demonstrate why looking for trends must be done with great caution (Table 5).

Table 5. Effect of ICD revisions on coding of sarcoma.

Diagnosis: Sarcoma

ICD Revision	ICD Code	ICD Category
ICD-7	199.2	Malignant Neoplasm of Other and Unspecified Sites with Primary Site not Indicated
ICD-8	199.1	Malignant Neoplasm Without Specification of Site, Other
ICD-9	171.9	Malignant Neoplasm of Connective and Other Soft Tissue, Site Unspecified

The general term sarcoma would be classified as ICD Code 199 under both ICD-7 and ICD-8 as a cancer, site unspecified. But under ICD-9 it becomes an ICD Code 171.9, Malignant Neoplasm of Connective and Other Soft Tissue, Site Unspecified. This means an increase in the number of connective tissue cancers that are reported in national statistics may not be because of any real increase in the number of soft tissue sarcomas but simply as a result of a change in coding convention, a change in the logic our nosologists use.

With all of this in mind, let us now examine the other two sources which suggest a relationship between dioxin and soft tissue sarcomas.

CASE REPORTS OF U.S. EMPLOYEES

Marilyn Fingerhut discussed the seven case reports: three had no documented exposures to the chemicals of interest, two of those with exposure were actually carcinomas, and two of the exposed had histologically-confirmed soft tissue sarcomas (this symposium). Both of these sarcomas were malignant fibrous histiocytomas. One died in 1978, the other in 1983. Thus, one death would be coded by ICD-8 rules, the other by ICD-9 conventions. Under ICD-8, a malignant fibrous histiocytoma is a skin cancer; under ICD-9, it is a malignant neoplasm of connective and other soft tissue.

EXCESS CONNECTIVE TISSUE CANCER

According to an unpublished EPA report, connective tissue cancer among white women in Midland County was reported to be about four times the national average between 1950 and 1978, with the greatest number in the latter years. On the surface this looks quite impressive, but if one looks a bit deeper a different picture presents itself. The fourfold increase represents a total of thirteen cases between 1950 and 1978, about one case every 2-plus years. Seven of the thirteen occurred between 1970 and 1978.

The Michigan Department of Public Health investigated and found no commonalities among the women which would suggest a single causative agent, including dioxin (21). What they did find was that a number of women moved to Midland County with pre-existing disease. They also found that most of the death certificates, while providing the histological diagnosis, made no mention of the site of the tumor. You will recall from the leiomyosarcoma example, that such a death is usually placed in the ICD category Malignant Neoplasm of Soft and Connective Tissue. As the Michigan Department of Public Health noted, if the site of the tumors had been listed, a number of these deaths may have been coded in alternate categories and the statistics for Midland County could have changed dramatically. In other words, the statistics for connective tissue cancer among white women may represent a coding artifact and not a true biological excess.

CONCLUSIONS

From an epidemiological perspective, the mortality statistic with the greatest validity is overall death. The next order of validity includes the major categories -- total malignancies, cardiovascular deaths, and accidents. Numbers or rates associated with the rarer forms of death must be looked upon with a certain amount of skepticism.

In U.S. workers with documented heavy exposures to 2,3,7,8-TCDD, overall deaths have been less than expected, well below what one could attribute to the "healthy worker effect" (usually observed because industrial employees are on average, for their age, healthier than others in the general population). Deaths due to cardiovascular disease and cancer have been at or below expectations. In other words, the mortality statistics with the greatest validity do not indicate that even relatively high levels of exposure to 2,3,7,8-TCDD produce any increase in mortality. In the Dow 2,4,5-T workers, a similar picture was seen (22). And among residents in Midland County, cancer mortality has been 15% below that of the state (23).

Having said all of this, I still want to mention that we do have two deaths from soft tissue sarcoma among the Monsanto and Dow trichlorophenol workforce. What caused these two deaths? I don't know. More research is needed to explore the possible association between relatively high-level exposures to dioxin or trichlorophenol and soft tissue sarcomas; but, to echo the comments of Marilyn Fingerhut, during the course of this research we must address both histological and nosological misclassifications that may produce under- or over-estimates of the true risks associated with exposures.

REFERENCES

1. Hardell, L. and Sandstrom, A. Case-Control Study: Soft-Tissue Sarcomas and Exposure to Phenoxyacetic Acids or Chlorophenols. British Journal of Cancer 39:711-717, 1979.

2. Eriksson, M., Hardell, L., Berg, N.O., Moller, T. and Alexson, O. Soft-Tissue Sarcomas and Exposure to Chemical Substances: A Case-Referent Study. British Journal of Industrial Medicine 38:27-33, 1981.

3. Zack, J.A. and Suskind, R.R. The Mortality Experience of Workers Exposed to Tetrachlorodibenzodioxin in a Trichlorophenol Process Accident. Journal of Occupational Medicine 22:11-14, 1980.

4. Cook, R.R., Townsend, J.C., Ott, M.G. and Silverstein, L.G. Mortality Experience of Employees Exposed to 2,3,7,8-Tetrachlorodibenzo-p-dioxin (TCDD). Journal of Occupational Medicine 22:530-532, 1980.

5. Cook, R.R. Dioxin, Chloracne, and Soft Tissue Sarcoma. Lancet 1:618-619, 1981.

6. Zack, J.A. and Gaffey, W.A. A Mortality Study of Workers Employed at The Monsanto Company Plant in Nitro, West Virginia. Tucker, R.E., Young, A.L. and Gray, A.P., Eds. Human and Environmental Risks of Chlorinated Dioxins and Related Compounds. New York: Plenum, 1983. pp. 575-591.

7. Riggan, W.B., Van Bruggen, J., Acquavella, J.F., Beaubier, J. and Mason, T.J., U.S. Cancer Mortality Rates and Trends, 1950-1979. Environmental Protection Agency, Washington, DC. Superintendent of Documents, U.S. Government Printing Office, No. 055-000-00234-7; January 24, 1984.

8. Cole, P. Direct Testimony of Dr. Philip Cole Before the Environmental Protection Agency of the United States of America In Re: The Dow Chemical Company, et al. FIFRA Docket Nos. 415, et al. (Exhibit 860) Appearance Date: November 6, 1980.

9. Rt. Hon. Peter Walker, M.P., Minister of Agriculture, Fisheries and Food, In: Advisory Committee on Pesticides: Report on Phenoxy Acid Herbicides. London: Ministry of Agriculture, Fisheries and Foods, February 7, 1983.

10. Cook, R.R. Soft Tissue Sarcomas: Clues and Caution. Tucker, R.E., Young, A.L. and Gray, A.P., Eds. Human and Environmental Risks of Chlorinated Dioxins and Related Compounds. New York: Plenum, 1983. pp. 613-618.

11. Smith, A.H., Fisher, D.O., Giles, H.J. and Pearce, N. The New Zealand Soft Tissue Sarcoma Case-Control Study: Interview Findings Concerning Phenoxyacetic Acid Exposure. Presented at 3rd International Symposium on Chlorinated Dioxins and Related Compounds, October 12-14, 1982, Salzburg, Austria.

12. Smith, A.H., Fisher, D.O., Pearce, N. and Teague, C.A. Do Agricultural Chemicals Cause Soft Tissue Sarcoma? Initial Findings of a Case-Control Study in New Zealand. Community Health Studies 6:114-119, 1982.

13. Milham, S., Jr. Herbicides, Occupation, and Cancer. Lancet 1:1464-1465, 1982.

14. Hardell, L. Malignant Lymphoma of Histiocytic Type and Exposure of Phenoxyacetic Acids and Chlorophenols. Lancet 1:55-56, 1979.

15. Eklund, G. Does Occupational Exposure to Chemical Pesticides Increase the Risk of Cancer? In: Weed and Plant Protection Conferences. Swedish University of Agricultural Sciences Research Information Centre, Uppsala, Sweden, 1983. pp. 6-12.

16. Manual of the International Statistical Classification of Diseases, Injuries and Causes of Death. 1965 Revision. Background. Geneva: World Health Organization, 1967. pp. VII-XXXIII.

17. Manual of the International Statistical Classification of Diseases, Injuries and Causes of Death. ICDA-8. Adapted for use in the United States by U.S. DHEW, Public Health Services, National Center for Health Statistics. Washington, DC: U.S. Government Printing Office (PHS Publication No. 1693), 1968.

18. National Cancer Institute. Monograph 41. Third National Cancer Survey: Incidence Data. DHEW Publication No. (NIH) 75-787. Washington, DC: U.S. Government Printing Office, 1975.

19. Manual of the International Statistical Classification of Diseases, Injuries and Causes of Death. 1955 Revision. Geneva: World Health Organization, 1957.

20. Manual of the International Statistical Classification of Diseases, Injuries and Causes of Death. 1975 Revision. Geneva: World Health Organization, 1977.

21. Evaluation of Soft and Connective Tissue Cancer Mortality Rates for Midland and Other Selected Michigan Counties Compared Nationally and Statewide. Lansing: Michigan Department of Public Health, 4 May 1983. Unpublished.

22. Ott, M.G., Holder, B.B. and Olson, R.D. A Mortality Analysis of Employees Engaged in the Manufacture of 2,4,5-Trichlorophenoxy-acetic Acid. Journal of Occupational Medicine 22:47-50, 1980.

23. Walker B. Michigan Department of Public Health News, 9 December 1982.

REPRODUCTIVE TOXICITY OF TETRACHLORODIBENZO-*p*-DIOXIN

Donald R. Mattison, M.D.
Maria S. Nightingale, M.S.
Ellen K. Silbergeld, Ph.D.

Section on Reproductive Toxicology
Pregnancy Research Branch
National Institute of Child Health and Human Development
National Institutes of Health
Bethesda, Maryland 20205

From Public Health Risks of the Dioxins, proceedings of a symposium held on October 19-20, 1983 at The Rockefeller University, New York City. Edited by William W. Lowrance. Published by William Kaufmann, Los Altos, California, 1984.

INTRODUCTION

The deliberations of this symposium are a modern version of a Greek tragedy. We are concerned about the adverse effects of xenobiotics on humankind and on the environment, yet many of these xenobiotics contribute to the quality of life we enjoy. TCDD (2,3,7,8-tetrachlorodibenzo-p-dioxin) is unique in that it neither contributes to the quality of life nor improves our standard of living. In many respects TCDD is the prototypical by-product: it is a by-product in the production of the herbicide 2,4,5-T and other chlorophenols and of the technology of a previous decade (1,2,7).

TCDD can also be considered a prototypical reproductive toxin (3-5), our focus for this paper. TCDD, like many other xenobiotics, has been investigated for distribution, metabolism, and toxicity in the adult animal (1,6-9). However, exploration of its reproductive effects have been minimal.

This overview of the reproductive toxicity of TCDD will begin by defining successful reproduction and reproductive toxicity. Next it will review the effects of TCDD on reproduction in experimental animals, including some data on TCDD's effects on the rodent and nonhuman primate ovary. Finally, it will review epidemiological studies of human exposure to chlorophenols contaminated with TCDD, focusing specifically on the reproductive endpoints explored by these studies.

CHARACTERISTICS OF REPRODUCTION

The reproductive system has many unique characteristics including intermittent function; therefore the timing of exposure to putative reproductive toxins is important in determining effect. In addition, reproductive function differs from that of other organ systems (3-5):

1. The reproductive system is the only organ system requiring interaction with tissue from another individual for successful functioning.

2. The reproductive system does not function continuously in the life cycle of the animal.
3. Reproductive system failure does not threaten survival of an individual, although it may have an impact on his or her lifestyle.

One complexity with respect to human reproductive function is that it involves sexuality. Reproduction goes beyond physiology and affects personalities. Altered fertility or reproductive function can therefore influence interactions with others as well as self-esteem. For these reasons, when we evaluate human populations, it is essential to deal with the physiological, toxicological, and emotional facets of reproduction.

Costs of reproduction. It is fortunate that reproduction is intermittent because the biologic cost of reproduction is high. For example, maintenance of optimal fertility in the female may be compromised by changes in physical state such as exercise (3). During gestation there are metabolic and physiological demands, and although most females adapt to stresses of pregnancy, there are situations in which some cannot. To our knowledge there is no evidence that exposure to TCDD or any of the polychlorinated dibenzodioxins (PCDD) or polychlorinated dibenzofurans (PCDF) preceding pregnancy alters adaptation and survival of the female in a subsequent pregnancy. However, there is evidence that prior exposure to some xenobiotics can increase maternal morbidity during pregnancy (4,10). It has also been suggested the PCDD or PCDF stored in adipose tissue may be mobilized during pregnancy and lactation to produce fetal and neonatal toxicity (18,19,35-38).

Regulation of reproduction. Even after parturition, offspring make metabolic demands on the mother and are an economic burden for the parents. Because of these and other stresses during and after gestation it is important that reproduction be regulated. Reproduction is controlled by various hormones which act on the reproductive system directly or indirectly through effects on the central nervous system (3-5). Animals use endogenous hormonal mechanisms to control breeding seasons and to modulate breeding efficiency during stress (21). Endogenous and exogenous hormones also regulate human reproduction. Other factors, such as age, nutrition, exercise, and economic status can influence reproductive function in complex ways (3-5). The fact that reproduction is an intermittently-expressed function creates problems for the epidemiological assessment of adverse reproductive effects in human populations (3-5).

Successful reproduction. Successful reproduction must be evaluated differently from hepatic, pulmonary, cardiovascular, or immunologic functions by at least three criteria:

1. The reproductive systems of two animals, or a human couple, must function effectively and efficiently when the production of offspring is desired (or appropriate for the life cycle or season).
2. Reproduction must lead to the formation of healthy, normal offspring in terms of both structure and function.
3. There must be suppression of the reproductive system, i.e., contraception, when reproduction is not desired or not appropriate for the season or life cycle. In animals, reproduction outside the normal breeding season or in time of stress (i.e., starvation or overcrowding) can have a profound adverse impact on the health of the female. Similar disastrous health complications apply to humans when inappropriate reproduction occurs.

Such interference with contraception is not a trivial consideration in modern society, as there are xenobiotics which can impair the effectiveness of oral contraceptives (11). Most of these compounds act by increasing the clearance of oral contraceptives, a combination of an estrogen and progestagen, or a progestagen alone. Oral contraceptives act either by blocking ovulation and/or by producing a uterine environment which does not support implantation of the fertilized egg. Increased contraceptive clearance can result from an increased rate of hepatic metabolism or decreased entero-hepatic circulation of the steroids. This is important at the present time because of the growing trend to use oral contraceptives with ever-decreasing doses of estrogen and progestagen. Xenobiotics documented to decrease contraceptive efficacy include rifampin, an antibiotic, and several anticonvulsants including phenytoin (11). Xenobiotics suspected of decreasing the effectiveness of oral contraceptives include any compound or exposure which increases hepatic clearance of exogenous estrogens or progestagens, for example, by induction of monooxygenases as reported for cigarette smoking (45). TCDD, an inducer of some hepatic monooxygenases and other enzymes, may function in a similar fashion to increase contraceptive failure (1,3,8,9). While there is one report of occupational exposure (not involving PCDDs or PCDFs) that has been associated with increased fertility, it is not known whether this is due to a healthy worker effect, impaired contraceptive effectiveness, or statistical variation in fertility rates (12). In one occupational study of PCDD exposure at Dow Chemical (34), the oral contraceptive failure rate appeared to be higher among wives of workers exposed to PCDD than controls (See Table 5).

Reproductive Toxicant. These considerations lead to the definition of reproductive toxicity. If successful reproduction is the production of healthy offspring during the appropriate season or time in the life cycle, a reproductive toxin is any compound that impairs fertility when offspring are desired or appropriate or that increases the production of structurally or functionally abnormal offspring. Alternatively, a reproductive toxin may increase fertility at an inappropriate time, in either the wrong season or time in the life cycle, by altering hormonal control of reproduction.

REPRODUCTIVE TOXICITY OF TCDD

Based on the general toxic effects of TCDD (1,6-9), adverse effects on mammalian reproduction would be expected from TCDD exposure (3-5). Its extreme potency as an acute toxin to adult animals may suggest even greater prenatal toxicity given the sensitivity of fetal systems to xenobiotics. Studies have demonstrated both teratogenicity and fetal toxicity, including cleft palate, cystic kidneys, subcutaneous edema, and intestinal hemorrhage in animals exposed in utero to low doses (13-17); there also has been damage to the immune system in the neonate (18). In addition, prefertilization exposure of the maternal organism produces embryotoxicity during a subsequent pregnancy (18,19,35-37).

Because TCDD can alter the activity of some hepatic and gonadal microsomal cytochrome P-450 dependent monooxygenases as well as other enzymes, exposure may alter the synthesis or metabolism of endogenous or exogenous sex steroids. Altered pharmacokinetic or pharmacodynamic properties of the sex steroids will cause changes in hormonal regulation of reproduction, including contraception by synthetic sex steroids.

TCDD may also be a genotoxin; this has been difficult to define, due to conflicting evidence of mutagenicity in some bacterial systems and carcinogenicity in mammals (1,38). Alternatively, TCDD may act as a tumor promoter in animals previously exposed to an initiator (40). Compounds which cause mutations or chromosomal aberrations in germ cells can increase the rate of birth defects and spontaneous abortions. However, at present, the putative genotoxicity of TCDD must be explored in greater detail before this hypothesized mechanism of reproductive toxicity is confirmed.

Male reproductive effects. Interference with male reproductive function may be exerted at the gonadal, hormonal, or central nervous system level (3-5). Animal experiments demonstrate that, at relatively high semi-chronic doses of TCDD (above 1 ug/kg), testicular atrophy is evident, along with destruction of the seminiferous tubules and spermatogenic cells (22-24). Generally in monkeys, pigs, cows, rats, mice and chickens there is a dose-dependent decrease in testicular weight following TCDD treatment (1). However, as observed for lethality, there are species differences in sensitivity of the male reproductive system to TCDD. For example, the guinea pig, a species sensitive to systemic toxicity from TCDD, is also more sensitive to testicular toxicity. When treated with TCDD guinea pigs experience a dose-dependent decrease in both absolute and relative testicular weight. In mice, although the absolute testicular weight decreases, the testicular to body weight ratio remains unchanged (22). In rats, both the absolute and relative testicular weight remain unchanged following treatment with low doses of TCDD (23); however, at higher doses testicular toxicity defined as depression of testicular heme synthesis and P-450 content occurs (47,48). In addition, rhesus monkeys treated with TCDD also have inhibition of spermatogenesis (24).

Changes in testis weight following TCDD treatment reflects changes in testicular structure. There are varying degrees of spermatogenic arrest in proportion to the effect on testicular weight (22-24). The seminiferous tubules appear to decrease in diameter and have empty lumens reflecting the inhibition of spermatogenesis. The remaining Sertoli cells, spermatogonia, and spermatocytes in the seminiferous tubules have abnormal morphology. These testicular alterations are thought to be a terminal event rather than a specific end organ toxicity (22,23). However, careful morphometric analysis of testis from surviving animals has not been reported. Recently, carefully designed pair feeding studies have suggested that the effects of TCDD on the testis are not due to inanition but instead to organ toxicity (see Peterson, this symposium). Despite reports demonstrating significant suppression of spermatogenesis, no alteration was observed in mouse fertility or sperm structure following treatment with mixtures containing TCDD (20,48). TCDD was found to be negative in standard assay for dominant lethality (38,49), and no increased incidence in stillbirth or birth defects was observed in rodents fathered by exposed males (48). These studies suggest TCDD is not a potent genotoxic agent in gonadal cells.

Although popular literature refers to increased malformations and infertility in men exposed to TCDD in herbicides, there is no experimental basis for these reports. Assessment of male fertility following occupational exposure has been indirectly reported in several studies; fertility does not appear to be altered (41). However, careful assessment for transient or permanent alterations in fertility or testicular function in exposed populations has not been reported (41,34).

TCDD is a potent inducer of some hepatic and testicular monooxygenases and other enzymes and thus it would be expected to alter testosterone levels in exposed animals by altering androgen pharmacokinetics. Several researchers have demonstrated that TCDD induces testicular aryl hydrocarbon hydroxylase (AHH) activity, a monooxygenase activity which probably reflects several enzymes (26,27) (Table 1).

Table 1. Murine hepatic and gonadal aryl hydrocarbon hydroxylase activity (pmoles/mg S-9 protein/min).

STRAIN		ORGAN	CONTROL	TCDD
DBA/2N	Male	Liver	139.0 ± 13	1019 ± 94
		Testis	6.5 ± 0.1	380 ± 26
	Female	Liver	67.0 ± 7.0	508 ± 8
		Ovary	3.0 ± 0.2	148 ± 9
C57BL/6N	Male	Liver	144.0 ± 8.0	129 ± 52
		Testis	5.1 ± 0.9	529 ± 22
	Female	Liver	67.0 ± 1.5	416 ± 56
		Ovary	3.3 ± 0.4	129 ± 11

Mice were treated with TCDD (20 ug/kg) 3 days before sacrifice; hepatic, ovarian and testicular 9000xg supernatant (S-9) was prepared and aryl hydrocarbon hydroxylase activity was measured as described in Reference 26.

Previous experiments in our lab have demonstrated a substantial increase in testicular AHH activity following treatment with TCDD. Lee et al. (42) have also demonstrated increased testicular and prostatic benzo(a)pyrene metabolism following TCDD treatment, and others have demonstrated substantial increases in hepatic metabolism of other xenobiotics (28-31,43). [Recent reports by Peterson (this symposium) have demonstrated decreased circulating levels of androgens in TCDD-exposed animals.] Interestingly, the increase in testicular AHH activity appears to occur along with a significant decrease in testicular P-450 content. At oral doses as low as 0.08 ug/kg, the activity of some hepatic androgen hydroxylases is decreased. However, this effect on sex steroid metabolism depends on the particular monooxygenase studied as well as the organ and sex of the animal (30,32,33,46,47,50,51).

Although TCDD is a potent mammalian toxin, its effects on male reproduction or reproductive outcome mediated through male exposure have been difficult to quantify. Carefully conducted experiments are needed utilizing pair feeding with more complete analyses of both spermatogenesis and hormonal control. Existing evidence from animal studies suggests that female reproductive function and embryonic development are the most sensitive to interference by TCDD.

Female reproductive effects. Female reproduction appears to be exquisitely sensitive to the adverse effects of TCDD. For example, although an oral dose of 0.01 ug/kg in adult rats (5 days/week x 13 weeks) produces no apparent somatic toxicity, the same dose produces significant alteration in reproductive function in a three generation study (25). Rats from F1 and F2 generations treated with TCDD at 0.01 ug/kg/day in food had significant decreases in fertility, litter size at birth, live born pups, pup survival and growth after birth. Offspring of rats treated at 0.001 ug/kg/day also had decreased pre- and postnatal survival (25).

TCDD has also been investigated for its ability to affect hormonal function in female experimental animals. At doses between 1 and 2 ug/kg in monkeys (43) and rodents (30,32,33,50,51), there are reports of altered hormone metabolism, including increases in the rate of 7-a-hydroxylation of testosterone and decreased catechol estrogen formation. At relatively high doses of 1 ug/kg/day for 13 weeks, changes in estrous cycling (including anovulation) have been described (23). Decreased numbers of corpora lutea were observed in ovaries from rats treated with TCDD at a dose of 2 ug/kg/day (54). Another indication of altered hormonal levels may be the reduced incidence of the estrogen-dependent tumors in uterus and mammary glands in exposed rodents (23).

Another endpoint of female gonadal toxicity is integrity of oocyte populations. We have found that several carcinogens which induce AHH activity are potent cytotoxic (genotoxic) agents that destroy small oocytes (3). Mice treated with TCDD show a gradual loss of small oocytes (Table 2), similar to the loss reported after exposure to benzo(a)pyrene, 3-methylcholanthrene, or dimethylbenzanthracene (3).

Table 2. Effect of TCDD on murine small oocyte number.

TREATMENT	SMALL OOCYTES/OVARY DBA/2N	(MEAN ± SD) C57BL/6N
I. Control	4290 ± 428	3610 ± 548
TCDD		
5 days	3885 ± 1228	3275 ± 940
11 days	3680 ± 799	2840 ± 502
II. Control	4334 ± 930	2320 ± 845
TCDD	2716 ± 828	2260 ± 1316
3-MC	629 ± 263	68 ± 120
TCDD + 3-MC	564 ± 368	72 ± 118

In experiment I, mice were treated with TCDD dissolved in dioxane (20 ug/kg) or dioxane (1 ul) on day 0. Mice treated with TCDD were sacrificed on day 5 or 11, and control mice were sacrificed on day 11 after treatment. In experiment II, mice were treated with TCDD (20 ug/kg) as in experiment I. Mice treated with 3-methylcholanthrene (3-MC) received 100 mg/kg ip in corn oil. Control mice received both corn oil (1 ml) and dioxane (1 ul). Mice in experiment II were sacrificed on day 13 following treatment. Ovaries were prepared and oocyte counts performed as described in Reference 52.

The effects of TCDD were not additive with 3-methylcholanthrene, raising the question as to mechanism of action. The effect of oocyte destruction is to shorten the reproductive lifespan and to produce early ovarian failure (52). TCDD exposure induces AHH activity in both liver and ovary (see Table 1), but the significance of this for functional integrity of ovarian cells and germ cell toxicity is not understood at present. However, TCDD may increase the capacity of the ovary to form genotoxic metabolites from xenobiotics. TCDD increases by twofold the ability of cynomologus monkey ovarian microsomes to form the 7,8- and 9,10-diol metabolites from benzo(a)pyrene (Table 3).

Table 3. Effect of intraovarian injection of TCDD on ovarian benzo(a)pyrene metabolism (pmole/mg microsomal protein/min).

METABOLITE	CONTROL	TCDD	TCDD/CONTROL RATIO
9,10-DIOL	.039	.071	1.8
4,5-DIOL	.058	.095	1.6
7,8-DIOL	.156	.317	2.0
1,6-DIONE	.444	.125	0.3
3,6-DIONE	.470	1.137	2.8
6,12-DIONE	.173	.395	2.3
9-phenol	.147	.288	2.0
3-phenol	.317	.609	1.9

Two adult cynomologus monkeys were treated with intraovarian injection of TCDD in dioxane (0.2 ug/ovary). Each animal served as its own control in that one ovary was treated with dioxane alone (10 ul) while the other ovary was treated with dioxane containing TCDD (20 ug/ml). The ovaries were removed 48 hrs later and microsomes prepared and utilized for benzo(a)pyrene metabolism studies as described in Reference 52.

The effects of TCDD on hepatic metabolism must also be examined for indirect effects on other endpoints of female reproduction (3). Uterine responsiveness to estrone is not affected by acute TCDD exposure (51); however, no studies have been undertaken during chronic exposure when altered receptor regulation might be expected.

Ovarian-uterine effects. Although the mechanism of TCDD impairment of female reproductive function is not completely understood, reproductive efficiency is impaired in female rodents treated with TCDD. Copulatory frequency of female rats treated with TCDD is decreased and the interbirth interval is increased (25); the number of corpora lutea and uterine weight are also decreased (19,23). The uterine endometrium in these TCDD-treated animals demonstrates a simple aglandular pattern consistent with impaired ovarian hormone production rather than the complex glandular structure observed with normal ovarian function (23). There is also a decrease in the incidence of spontaneous uterine tumors. All these observations suggest either a decrease in the level of estrogen and progesterone reaching the uterus or an impaired response of the uterus to these hormones. Confirmation of the first hypothesis has been suggested by serum hormone measurements in rhesus monkeys which demonstrate a decrease in circulating levels of estrogen and progesterone (44). However, recent studies demonstrate impaired hepatic estrogen metabolism in TCDD-treated pregnant rats, and other studies have been unable to demonstrate an alteration in the uterotropic action of estrone in TCDD-treated ovariectomized rats, suggesting that TCDD does not alter the clearance of estrogens (19,30). Although there is controversy concerning the effects of TCDD on steroid hormone pharmaco_kinetics_, no experiments have explored the effects of TCDD on steroid hormone pharmaco_dynamics_.

In rhesus monkeys, TCDD feeding decreases the circulating levels of estrogen and progesterone and results in abnormal menstrual cycles (43). These observations suggest that TCDD decreases the production and/or increases the clearance of estrogen and progesterone in nonhuman primates. The effect of occupational or environmental exposure to TCDD on human or nonhuman primate hypothalamic-pituitary-ovarian-uterine function is unknown.

Fetal and Neonatal Effects. TCDD appears to be one of the most potent teratogens and embryo/fetal toxins described in rodents (25). At doses as low as 0.1 ug/kg during mid-gestation, offspring are produced with skeletal anomalies, cleft palate, and malformed kidneys. In addition, prenatal exposure to TCDD is associated with lower birth weight, decreased rate of survival, and immunological defects (18,19). Gestational or pregestational exposure of females to TCDD also increases the rate of spontaneous abortion and/or resorption. Increased fetal death and abortion have also been reported in rhesus monkeys exposed to TCDD at doses as low as 1 ug/kg (McNulty, this conference).

Toxicity in the offspring also results from exposure to TCDD via breast milk from nursing mothers (35,53). Few measurements of human breast milk concentrations of TCDD have been reported; however, in a recent review, Jensen found TCDD levels in the ppb range in breast milk of presumably non-occupationally exposed women (35). Effects from this type of exposure would be expected to be severe, based on the clinical experience in Yusho patients with the dibenzofurans, which are structurally and toxicologically similar. Human exposure to the dibenzofurans is associated with hirsutism, persistent hyperpigmentation, and abnormal development in children (44). In addition, women exposed to dibenzofurans or polychlorinated biphenyls report abnormal menses (3,7,44).

Fetal sensitivity to TCDD may relate to the increased rates of spontaneous abortion in exposed animals, decrease in birthweight, and intestinal hemorrhage in livebirth (54). These effects can be produced in offspring in animals exposed before conception. However, these findings do not suggest long-term impairment of the maternal reproductive system, since it is likely that measurable levels of TCDD were still present in plasma and tissues of exposed females during gestation (55). It should also be noted that some reproductive effects are observed at relatively low doses of TCDD, at least compared with those associated with somatic toxicity and significantly increased rates of tumor formation.

EPIDEMIOLOGICAL DATA

We will briefly review the reproductive toxicity reported in epidemiological studies of five human populations exposed to chlorophenols which were contaminated with TCDD (56).

Seveso. Following the explosion of the ICMESA trichlorophenol plant in 1976 in Seveso, Italy, several cohorts of exposed and presumably nonexposed persons have been followed. It is difficult to determine if any acute effects on reproduction occurred, since fertility was not monitored or evaluated and therapeutic abortions were offered to pregnant women living in the area of highest contamination. Although chromosomal abnormalities were detected in fetal tissue from abortuses (39), the fetuses were thought to be normal (60). Ultrastructural changes were observed in the placentas from women who received therapeutic abortions following exposure to TCDD in Seveso (59). Follow-up studies of the Seveso populations since 1976 have reported some increases in certain birth defects in children born to exposed women: anencephaly, spina bifida, hydrocephaly, hypospadias, polydactyly, and hemangiomas (8). However, the absolute numbers of events recorded are very low, and information on control or background rates of such defects in Italy is unclear. In the same period, 1976-1978, a short term increase in the rate of spontaneous abortion has been reported, but circumstances surrounding abortion in Italy make it difficult to ascertain true rates (Table 4).

Table 4. Number and rates of spontaneous abortions in Seveso by quarter.[a]

		1976		1977				1978	
		3°	4°	1°	2°	3°	4°	1°	2°
Zone B	A	3	4	5	8	10	4	3	3
	D	27	18	29	28	32	29	21	23
	R	11.1	22.2	17.2[b]	28.5[b]	31.2	13.7	14.3	8.6
Zone R	A	19	17	15	17	16	20	17	22
	D	138	104	118	135	140	144	118	102
	R	13.7	16.3	12.7	12.5	11.4	13.8	14.4	21.5
Zone Non-ABR	A	74	94	119	81	67	99	78	70
	D	670	632	713	621	634	691	596	555
	R	11.0	14.8	16.6	13.0	10.5	14.3	13.1	12.6

[a]Report on the obstetric situation to the International Steering Committee, 4th Meeting, November, 1980; from Reference 8.

[b]Cumulative values for the period 4°, 1976 through 3°, 1977: Zone B=25.2, Zone R=13.1, and Zone Non-ABR=13.9.

A=abortion, D=delivery, and R=rate.

Currently there is no information either on reproductive status of persons exposed prior to sexual maturity or on total reproductive history of exposed adults, and no attempt is being made to collect data on reproductive function.

Herbicide spray studies. Several areas have been investigated to determine adverse effects on the general population from exposure to TCDD in phenoxyacetic acid herbicides (2,4,5-T). The most carefully studied areas are in New Zealand (56), Alsea, Oregon (57), and Arkansas (58). In all instances, information is limited to the time of herbicide spraying with little or no data on environmental concentrations or the extent of product contamination by TCDD. If these exposures were to a combination of 2,4,5-T and TCDD, this increases the difficulty of determining an etiologic role specific to TCDD. Although these studies suggest an increase in skeletal anomalies, information on background rates is not established. With respect to spontaneous abortions, the Alsea study reported an increase during only one period of spraying. The CDC study failed to find an increase in the rate of specific birth defects or cleft palate in children born to women residing in areas with heavy agricultural applications of 2,4,5-T.

Hexachlorophene. This germicide may contain dioxins, and in the past this contamination may have been greater. One study of nurses exposed occupationally to hexachlorophene reported a higher rate of birth defects in their children compared with those of unexposed nurses (9). However, the rate of malformations in the exposed nurses' children is unusually low (about 5%, 25 out of 487 births) while in the control group, no malformations were reported in 227 births. Moreover, hexachlorophene is itself known to be fetotoxic and possibly teratogenic, so that it is not possible to ascribe these effects entirely to TCDD exposure.

Dow/Midland County. Two studies have examined the reproductive history of workers and their families at the Midland, Michigan plant of Dow Chemical--one of the largest volume producers of trichlorophenols and 2,4,5-T in the world. The first study was conducted by Dow (34); the second study was conducted by the Michigan State Department of Public Health and is available only as a report (61). The study populations, while overlapping, are not identical; Dow's study covered approximately 400 families drawn from 930 male employees of the Midland plant, and the Michigan study surveyed hospital records for Midland County, which included those of Dow employees and others. Exposure was calculated on the basis of work description and tenure for the Dow study and on the basis of residence for the Michigan study.

The Dow study explored effects of male occupational exposure to chlorophenols (which are contaminated with varying amounts of TCDD) on the reproductive performance of their wives (34) (Table 5).

Table 5. Effect of paternal PCDD exposure on female reproductive function.

	PCDD EXPOSED HUSBAND (737)		UNEXPOSED HUSBAND (2031)	
	n	(%)	n	(%)
Maternal Age at Conception				
<18	8	(1)	113	(6)
18-30	539	(73)	1675	(82)
>30	190	(26)	243	(12)
Gravidity				
1	96	(13)	603	(30)
2-3	297	(40)	890	(44)
>4	344	(47)	538	(26)
Contraception at Time of Pregnancy				
None	658	(89)	1884	(93)
Oral Contraceptives	14	(2)	12	(<1)
Intrauterine Device	5	(<1)	7	(<1)
Other	60	(8)	128	(6)
Maternal Health Problems During Pregnancy				
None	482	(65)	1447	(71)
Any	255	(35)	584	(29)
Maternal Health Treatment During Pregnancy				
None	473	(64)	1382	(68)
Any	264	(36)	649	(32)

Table 5. Continued.

	PCDD EXPOSED HUSBAND (737)		UNEXPOSED HUSBAND (2031)	
	n	(%)	n	(%)
Complications During Labor and Delivery				
None	456	(72)	1278	(72)
Any	181	(28)	507	(28)
Pregnancy Outcome				
Spontaneous Abortion	85	(12)	213	(10)
Rate/1000 Live Births	133		119	
Still Birth	15	(2)	33	(2)
Rate/1000 Live Births	24		19	
Live Birth	637	(86)	1785	(88)

Modified from Townsend (34).

Although the wives of men occupationally exposed to any dioxin were slightly more likely to have medical problems (unspecified) and to require medication during pregnancy, they were also older and of higher gravidity than the controls. Also more wives of PCDD and PCDF exposed men were contracepting at the time of pregnancy than were the wives of the non-exposed men, suggesting a high contraceptive failure rate among these women. This is surprising in view of their greater age and higher gravidity, factors expected to decrease the possibility that pregnancy occurred because the women forgot to take their oral contraceptives.

No difference in complications during labor and delivery was observed between the two groups. However, there did appear to be small differences in pregnancy outcome: spontaneous abortion and stillbirth rate were higher among wives of exposed men. The Dow study reported no significant changes in birth defects, infertility, or spontaneous abortion. The Michigan Department of Public Health study reported a cluster of cases of spina bifida and cleft palate in Midland County; however, these findings are difficult to interpret because of changes in case ascertainment during the study. Another objection to the Michigan state study is that other clusters of these defects were found in other Michigan counties far from Midland. However, this may not be relevant since there may be other causes of these physical defects; for example, other areas of Michigan are contaminated by agents with toxicological similarities to the dioxins, such as the polybrominated biphenyls.

Agent Orange and Vietnam. Two populations exposed to military use of herbicides have been studied, although neither rigorously. Families of Vietnam veterans exposed to Agent Orange have reported a range of reproductive effects, including impotence, loss of libido, infertility, birth defects, and neonatal deaths (41). If these effects have occurred they could be associated with male-mediated exposures; however, on the basis of the experimental data reviewed in this paper, it is possible that male sexual dysfunction could occur. There are no experimental data supporting paternally transmitted birth defects resulting from TCDD exposure (20,38). In Vietnamese populations exposed to the herbicide, a high rate of spontaneous abortions and birth defects has been reported (56); however, the data are primarily case reports with little systematic collection of data. Also, these populations have endured numerous other hardships in the course of the war, which makes it difficult to select TCDD as a primary cause of reproductive problems. Nevertheless, this population should be studied further and attempts to characterize specific exposure groups should be undertaken.

CONCLUSIONS

There is a discrepancy between the reproductive effects reported from TCDD (particularly at low dose) in animals and the endpoints usually examined in humans. For example, although fertility and preconceptional stages of reproduction appear to be impaired in animals, these endpoints have not been evaluated in exposed human populations. It appears that the endocrine system is quite sensitive to TCDD exposure and that alterations in female reproductive function may provide the earliest indications of toxicity. However, a substantial amount of work is required before the assessment of reproductive toxicity from TCDD or any other xenobiotic can be conducted critically. There are enormous gaps in our knowledge concerning the epidemiology of most reproductive endpoints. At the present time only a few adequate animal models of reproductive toxicity exist. We need to recognize and respond to these deficits by stimulating research in reproductive biology, reproductive epidemiology, and reproductive toxicology.

REFERENCES

1. Poland, A. and Knutson, J.C. 2,3,7,8-Tetrachlorodibenzo-p-dioxin and related halogenated aromatic hydrocarbons: Examination of the mechanism of toxicity. Annual Review of Pharmacology and Toxicology 22:517-554, 1982.

2. Holmstedt, B. Prolegomena to Seveso. Archives of Toxicology 44:211-230, 1980.

3. Mattison, D.R., ed. Reproductive Toxicology. Alan R. Liss, New York, 1983.

4. Barlow, S.M. and Sullivan, F.M. Reproductive Hazards of Industrial Chemicals. Academic Press, New York, 1982.

5. Vouk, V.B. and Sheehan, P.J., eds. Methods for Assessing the Effects of Chemicals on Reproductive Functions. Wiley, New York, 1983.

6. Tucker, R.E., Young, A.L. and Gray, A.P., eds. Human and Environmental Risks of Chlorinated Dioxins and Related Compounds. Plenum Press, New York, 1983.

7. Kimbrough, R.D., ed. Halogenated Biphenyls, Terphenyls, Naphthalenes, Dibenzodioxins, and Related Products. Elsevier, New York, 1980.

8. Garattini, S., ed. TCDD toxicology with particular reference to Seveso. Drug Metabolism Reviews 13: 345-469, 1982.

9. Nicholson, W.J. and Moore, J.A. Health effects of halogenated aromatic hydrocarbons. Annals of the New York Academy of Sciences 320:1-730, 1979.

10. Mattison, D.R., Gates, A.H., Leonard, A., Wide, M., Hemminki, K. and Copius Peereboom-Stegeman, J.H.J. Female reproductive system. In: Reproductive and Developmental Toxicity of Metals, T. W. Clarkson, G. Nordberg, and P. Sager, eds., pp. 41-91. Plenum Press, New York, 1983.

11. Breckenridge, A.M., Back, D.J. and Orme, M. Interactions between oral contraceptives and other drugs. Pharmacology and Therapeutics 7:617-626, 1979.

12. Levine, R.J., Symons, M.J., Balogh, S.A., Arndt, D.M., Kaswandik, N.T. and Gentile, J.W. A method for monitoring the fertility of workers. 1. Method and pilot studies. Journal of Occupational Medicine 22:781-791, 1980.

13. Courtney, K.D. and Moore, J.A. Teratology studies with 2,4,5-trichlorophenoxyacetic acid and 2,3,7,8-tetrachlorodibenzo-p-dioxin. Toxicology and Applied Pharmacology 20:396-403, 1971.

14. Courtney, K.D. Mouse teratology studies with chlorodibenzo-p-dioxins. Bulletin of Environmental Contamination and Toxicology 16:674-681, 1976.

15. Giavini, E., Prati, M. and Vismara, C. Rabbit teratology study with 2,3,7,8-tetrachlorodibenzo-p-dioxin. Environmental Research 27:74-78, 1982.

16. Neubert, D., Zens, P., Rothenwallner, A. and Merker, H.J. A survey of the embryotoxic effects of TCDD in mammalian species. Environmental Health Perspectives 5:67-79, 1973.

17. Kociba, R.J. and Schwetz, B.A. Toxicity of 2,3,7,8-tetrachlorodibenzo-p-dioxin (TCDD). Drug Metabolism Reviews 13:387-406, 1982.

18. Vos, J.G. and Moore, J.A. Suppression of cellular immunity in rats and mice by maternal treatment with 2,3,7,8-tetrachlorodibenzo-p-dioxin. International Archives of Allergy and Applied Immunology 47:777-794, 1974.

19. Giavini, E., Prati, M. and Vismara, C. Embryotoxic effects of 2,3,7,8-tetrachlorodibenzo-p-dioxin administered to female rats before mating. Environmental Research 31:105-110, 1983.

20. Lamb, J.C., Marks, T.A., Gladen, B.C., Allen, J.W. and Moore, J.A. Male fertility, sister chromatid exchange, and germ cell toxicity following exposure to mixtures of chlorinated phenoxy acids containing 2,3,7,8-tetrachlorodibenzo-p-dioxin. Journal of Toxicology and Environmental Health 8:825-834, 1981.

21. Van Tienhoven, A. Reproductive Physiology of Vertebrates. W.B. Saunders, Philadelphia, 1968.

22. McConnell, E.E., Moore, J.A., Haseman, J.K. and Harris, M.W. The comparative toxicity of chlorinated dibenzo-p-dioxins in mice and guinea pigs. Toxicology and Applied Pharmacology 44:335-356, 1978.

23. Kociba, R.J., Keeler, P.A., Park, C.N. and Gehring, P.J. 2,3,7,8-tetrachlorodibenzo-p-dioxin (TCDD): results of a 13-week oral toxicity study in rats. Toxicology and Applied Pharmacology 35:553-574, 1976.

24. Norback, D.H. and Allen, J.R. Biological responses of the nonhuman primate, chicken, and rat to chlorinated dibenzo-p-dioxin ingestion. Environmental Health Perspectives 5:233-240, 1973.

25. Murray, F.J., Smith, F.A., Nitschke, K.D., Humiston, C.G., Kociba, R.J. and Schwetz, B.A. Three-generation reproduction study of rats given 2,3,7,8-tetrachlorodibenzo-p-dioxin (TCDD) in the diet. Toxicology and Applied Pharmacology 50:241-252, 1979.

26. Mattison, D.R. and Thorgeirsson, S.S. Gonadal aryl hydrocarbon hydroxylase in rats and mice. Cancer Research 38:1368-1373, 1978.

27. Nagayama, J. and Lee, I.P. Comparison of benzo(a)pyrene metabolism by testicular homogenate and the isolated perfused testis of rat following 2,3,7,8-tetrachlorodibenzo-p-dioxin treatment. Archives of Toxicology 51:121-130, 1982.

28. Aitio, A. and Parkki, M.G. Organ specific induction of drug metabolizing enzymes by 2,3,7,8-tetrachlorodibenzo-p-dioxin in the rat. Toxicology and Applied Pharmacology 44:107-114, 1978.

29. Goldstein, J.A. The structure-activity relationships of halogenated biphenyls as enzyme inducers. Annals of the New York Academy of Sciences 320:164-178, 1979.

30. Shiverick, K.T. and Muther, T.F. 2,3,7,8-tetrachlorodibenzo-p-dioxin (TCDD) effects on hepatic microsomal steroid metabolism and serum estradiol of pregnant rats. Biochemical Pharmacology 32:991-995, 1983.

31. Hook, G.E., Orton, T.C., Moore, J.A. and Lucier, G.W. 2,3,7,8-tetrachlorodibenzo-p-dioxin-induced changes in the hydroxylation of biphenyl by rat liver. Biochemical Pharmacology 24:335-340, 1975.

32. Gustafsson, J.A. and Ingelman-Sandberg, M. Changes in steroid hormone metabolism in rat liver microsomes following administration of 2,3,7,8-tetrachlorodibenzo-p-dioxin (TCDD). Biochemical Pharmacology 28:497-99, 1979.

33. Nienstedt, W., Parkki, M., Uotila, P. and Aitio, A. Effect of 2,3,7,8-tetrachlorodibenzo-p-dioxin on the hepatic metabolism of testosterone in the rat. Toxicology 13:233-236, 1979.

34. Townsend, J.C., Bodner, K.M., Van Peenen, P.F.D., Olson, R.D. and Cook, R.R. Survey of reproductive events of wives of employees exposed to chlorinated dioxins. American Journal of Epidemiology 115:659-713, 1982.

35. Jensen, A.A. Chemical contaminants in human breast milk. Residue Reviews 89:1-128, 1983.

36. Luster, M.I., Boorman, G.A., Dean, J.H., Harris, M.W., Luebke, R.W., Padarathsingh, M.L. and Moore, J.A. Examination of bone marrow, immunologic parameters and host susceptibility following pre- and postnatal exposure to 2,3,7,8-tetrachlorodibenzo-p-dioxin (TCDD). International Journal of Immunopharmacology 2:301-310, 1980.

37. Moore, J.A., Gupta, B.N., Zinkl, J.G. and Vos, J.G. Postnatal effects of maternal exposure to 2,3,7,8-tetrachlorodibenzo-p-dioxin (TCDD). Environmental Health Perspectives 5:81-85, 1973.

38. Wassom, J.S., Huff, J.E. and Loprieno, N. A review of the genetic toxicology of chlorinated dibenzo-p-dioxins. Mutation Research 47:141-160, 1977/1978.

39. Tenchini, M.L., Crimaudo, C., Pacchetti, G., Mottura, A., Agosti, S. and De Carli, L. A comparative cytogenetic study on cases of induced abortions in TCDD exposed and nonexposed women. Environmental Mutagenesis 5:73-85, 1983.

40. Pitot, H.C., Goldsworthy, T., Campbell, H.A. and Poland, A. Quantitative evaluation of the promotion by 2,3,7,8-tetrachlorodibenzo-p-dioxin of hepatocarcinogenesis from diethylnitrosamine. Cancer Research 40:3616-3620, 1980.

41. Bogen, G. Symptoms in Vietnam Veterans exposed to Agent Orange. Journal of the American Medical Association 242 (22):2391, 1979.

42. Lee, I.P. and Suzuki, K. Induction of aryl hydrocarbon hydroxylase activity in the rat prostate glands by 2,3,7,8-tetrachlorodibenzo-p-dioxin. Journal of Pharmacology and Experimental Therapeutics 215:601-605, 1980.

43. Barsotti, D.A., Abrahramson, L.J. and Allen, J.R. Hormonal alterations in female rhesus monkeys fed diets containing 2,3,7,8-tetrachlorodibenzo-p-dioxin. Bulletin of Environmental Contamination and Toxicology 21:463-469, 1979.

44. Hirayama, C. Clinical aspects of PCB poisoning. In: PCB Poisoning, K. Higuchi, ed., pp. 87-104. Academic Press, New York, 1976.

45. Mattison, D.R. The effects of smoking on fertility from gametogenesis to implantation. Environmental Research 28:410-433, 1982.

46. Tofilon, P.J. and Piper, W.N. 2,3,7,8-tetrachloro-dibenzo-p-dioxin mediated depression of rat testicular heme synthesis and microsomal cyctochrome P-450. Biochemical Pharmacology 31:3663-3666, 1982.

47. Tofilon, P.J., Peters, P.G., Clement, R.P., Hardwicke, D.M. and Piper, W.N. Depressed guinea pig testicular microsomal cytochrome P-450 content by 2,3,7,8-tetrachlorodibenzo-p-dioxin. Life Sciences 27:871-876, 1980.

48. Lamb, J.C., Moore, J.A., Marks, T.A. and Haseman, J.K. Development and viability of offspring of male mice treated with chlorinated phenoxy acids and 2,3,7,8-tetrachlorodibenzo-p-dioxin. Journal of Toxicology and Environmental Health 8:835-844, 1981.

49. Khera, K.S. and Ruddick, J.A. Polychlorodibenzo-p-dioxins: perinatal effects and the dominant lethal test in Wistar rats. Advances in Chemistry Series 120:70-84, 1973.

50. Yoshihara, S., Nogata, K., Wader, I., Yoshimura, H., Kuroki, H. and Masuda, Y. A unique change of steroid metabolism in rat liver microsomes induced with highly toxic polychlorinated biphenyl (PCB) and polychlorinated dibenzofuran (PCDF). Journal of Pharmacobio-dynamics 5:994-1004, 1982.

51. Shiverick, K.T. and Muther, T.F. Effects of 2,3,7,8-tetrachloro-dibenzo-p-dioxin on serum concentrations and the uterotrophic actions of exogenous estrone in rats. Toxicology and Applied Pharmacology 65:170-176, 1982.

52. Mattison, D.R., Nightingale, M.S., Tukizawa, K., Silbergeld, E., Yagi, H. and Jerina, D. Benzo(a)pyrene reproductive toxicity and ovarian metabolism. In: Extrahepatic Drug Metabolism and Chemical Carcinogenesis, J. Rydstrom, J. Montelius, and M. Bengtsson, eds., pp. 337-350. Elsevier, New York, 1983.

53. Jensen, D.J. and Hummel, R.A. Secretion of TCDD in milk and cream following the feeding of TCDD to lactating dairy cows. Bulletin of Environmental Contamination and Toxicology 29:440-446, 1982.

54. Smith, F.A., Schwetz, B.A. and Nitschke, K.D. Teratogenicity of 2,3,7,8-tetrachlorodibenzo-p-dioxin in CF-1 mice. Toxicology and Applied Pharmacology 38:517-523, 1976.

55. Berry, D.L., Slaga, T.J., Wilson, N.M., Zachariah, P.K., Namkung, M.J., Brachen, W.M. and Juchau, M.R. Transplacental induction of mixed-function oxygenases in extra-hepatic tissues by 2,3,7,8-tetrachlorodibenzo-p-dioxin. Biochemical Pharmacology 26:1383-1388, 1977.

56. Hay, A. The Chemical Scythe: Lessons of 2,4,5-T and Dioxin. Plenum Press, New York, 1982.

57. Hanify, J.A., Metcalf, P., Nobbs, C.L. and Worsley, K.J. Aerial spraying of 2,4,5-T and human birth malformations: an epidemiological investigation. Science 212:349-351, 1981.

58. Nelson, C.J., Holson, J.F., Green, H.G. and Gaylor, D.W. Retrospective study of the relationship between agricultural use of 2,4,5-T and cleft palate occurrence in Arkansas. Teratology 19:377-384, 1979.

59. Remotti, G., DeVirgiliis, G., Bianco, V. and Candiani, G.B. The morphology of early trophoblast after dioxin poisoning in the Seveso area. Placenta 2:53-62, 1981.

60. Rehder, H., Sanchioni, L., Cefis, F. and Gropp, A. Pathologisch-embryologische Untersuchungen and Abortsusfallen in zusammenhang mit dem Seveso-Ungluck. Schweizerische Medizinische Wochenschrift 108:1617-1625, 1978.

61. Michigan Department of Public Health. Evaluation of Congenital Malformation Rates for Midland and Other Selected Michigan Counties Compared Nationally and Statewide, 1970-1981. Unpublished report, May 4, 1983.

FETOCIDAL AND TERATOGENIC ACTIONS OF TCDD

Wilbur P. McNulty, M.D.

Oregon Regional Primate Research Center
Beaverton, Oregon 97006

From Public Health Risks of the Dioxins, proceedings of a symposium held on October 19-20, 1983 at The Rockefeller University, New York City. Edited by William W. Lowrance. Published by William Kaufmann, Los Altos, California.

In this presentation, I will review briefly the limited published work on the effects of 2,3,7,8-tetrachlorodibenzo-p-dioxin (TCDD) on animal fetuses, comment on the usefulness and limitations of experimental findings for evaluation of human risks, discuss the suitability of using nonhuman primates for such investigations, and propose some future research.

Administered orally at a dose of 1 to a few ug/kg of body weight to pregnant females, TCDD has caused fetal loss or abortion in mice (1-3), rats (4), rabbits (5), and rhesus macaques (6). When the chemical was given orally on days 6-15 of gestation, similar or somewhat smaller doses induced cleft palate and dilated renal pelves in mice (1-3) and similar renal lesions in rats (4) and rabbits (5). The compound also produced cardiovascular malformations in chick embryos (7) and cranial anomalies in rainbow trout fry (8). The teratogenicity of TCDD for rhesus macaques has not yet been satisfactorily evaluated.

Continuous oral intake of TCDD at doses greater than 0.001 ug/kg/day diminished fertility and increased fetal loss in a three-generation feeding experiment with rats (9), and continuous oral intake of food containing 500 or 50 parts per trillion (ppt) of TCDD (0.02 or 0.002 ug/kg/day) for 6 months caused cyclic hormonal and menstrual irregularities, reduced fertility, and increased frequency of abortions in rhesus macaques (10,11). A high abortion rate was seen in mares environmentally exposed to TCDD (dose unknown) in Missouri horse arenas treated with waste oil containing TCDD (12).

In both the short- and long-term rodent experiments, fetocidal and teratogenic effects occurred at doses that did not cause observable signs of toxicity in the dams. On the other hand, the dose (1 ug/kg) or food level (50 or 500 ppt) that caused frequent abortions in rhesus macaques was sometimes also toxic for the mothers (6,10,11), although in the single-dose exposures, overt maternal toxic signs did not appear until after the abortions (Table 1).

Table 1. Abortions and maternal toxicity after administration of 2,3,7,8-tetrachlorodibenzo-p-dioxin (TCDD) to rhesus macaques in early pregnancy.

Dose (ug/kg body weight)		Abortions[a]	Days to abortion after dose[b]	Maternal toxicity	Days to toxicity after dose	Maternal death	Days to death after dose
Vehicle	0[c]	3/13	118 (1)	0/12	-	0/12	-
TCDD	0.2[c]	1/4	? (0)	0/4	-	0/4	-
	1.0[d]	13/16	20-60(9)	8/16	67-207	3/16	130-297
	5.0[c]	2/2	27-28(2)	2/2	30-61	2/2	65-116

[a]Pregnancy confirmed by products of conception or serum progesterone on days 20-22.

[b]Numbers of recoveries of products of conception in parentheses.

[c]Divided in nine intragastric doses in corn oil, days 20 to 40 post conception.

[d]Four animals given nine divided doses and twelve given a single dose, days 20 to 40 post conception.

Table 1 summarizes the results of treatment of rhesus monkeys, whose pregnancy was hormonally established before treatment, at three dose levels given from the 20th to the 40th day of pregnancy. Some of these were given in divided doses throughout the 20 days, while some were given in single doses at 20, 25, 30 and 40 days. This is the sensitive period of organogenesis; for example, thalidomide is most effective between the 23rd and the 29th or 30th day. In the first column, there were 13 of 16 abortions in the conglomerate group treated at 1 microgram per kilogram. When compared to 3 abortions out of 12 untreated monkeys, which is close to the expected historical frequency of abortion in rhesus monkeys of 15% to 20%, the difference is significant at less than 0.01 by the Fisher's exact test. At this one microgram per kilogram level, eight of the mothers showed evidence of toxicity, primarily as chloracne and weight loss, and three of these subsequently died. These signs appeared sometime after the abortion and the deaths occurred as late as eight months after the single dose. The abortion generally occurred from 10 to 20 days after the dose was given.

In the very limited investigations to date, the teratogenic potential of TCDD for rhesus macaques could not be assessed because the spontaneously aborted fetuses either were not recovered or were found in a macerated, fragmented condition that precluded anatomical study. The few surviving fetuses, which were surgically obtained live just before term, had no malformations (6). Testing of smaller doses in rhesus monkeys, which might distinguish direct adverse effects on the fetus from indirect action through deleterious effects on the mother, has not been done with enough animals to draw conclusions. However, it should be noted that in the short-term monkey tests, the fetocidal dose of TCDD was given when maintenance of pregnancy no longer depended on maternal ovarian progesterone (13)--a steroid hormone whose metabolism might be disturbed by the potent enzyme-inductive properties of TCDD.

TWO LIMITATIONS OF EXPERIMENTAL FINDINGS

The dilemmas encountered in designing experiments on reproductive effects in animals for the purpose of extrapolation to humans are familiar. They are similar to those encountered in research on experimental carcinogenesis.

1. To obtain conclusive results with manageable numbers of animals, one must use enough toxicant to produce frequent effects, often in the presence of a high "spontaneous" background incidence. But much lower environmental levels of such compounds, which might cause even infrequent adverse effects in human populations, would not be socially tolerable. Since the mechanisms of action of fetocidal or teratogenic agents are scarcely known in general, and those of TCDD are not known at all, there is little basis for a mathematical model of a dose-response relationship. Extrapolation to very low levels of exposure from the limited range of dose-response data that can be obtained in the laboratory is highly speculative.

The effective doses used in the short-term experiments in all species tested--of the order of 1 ug/kg of body weight--are well in excess of what might reasonably be expected from environmental (though not industrial) exposure. For example, a pregnant woman must eat a pound of meat containing 10% fat contaminated with TCDD at 140,000 ppt every day for a week to accumulate a dose of 1 ug/kg. This is a level of TCDD roughly 30,000 times greater than that reported for even an occasional sample of beef fat (14).

The excess is less for continuous dietary intake. Daily consumption of 50 g of fat containing 2,000 ppt of TCDD would provide a daily intake of TCDD equivalent to that reported by Allen (11) to decrease fertility and increase the frequency of abortions in rhesus macaques. This level is still about 500 times greater than the maximum reported for any beef fat samples.

2. Whether data on reproductive and fetal effects in one species are applicable to another species is questionable. Even the somatic sensitivities to TCDD and the spectra of pathologic lesions produced are quite different among laboratory species that have been tested. Also, since the normal reproductive strategies for timely mating, hormonal support of pregnancy, and placental nourishment of fetuses vary widely, an interspecies extrapolation of the reproductive effects of TCDD is unreliable.

SUITABILITY OF NONHUMAN PRIMATES

Old World nonhuman primates have female reproductive attributes closely resembling those of women. The cyclic hypothalamic-hypophyseal-ovarian hormonal events are nearly identical in pattern; the cyclic changes in the lining of the uterus are closely similar; and the structures of the placentas are almost the same. For these reasons, nonhuman primates--rhesus macaques in particular--are in my opinion the best experimental models for the effects of TCDD on human reproduction.

But monkeys are expensive and in short supply, and their husbandry is complicated. They are best used for qualitative investigations of the mechanisms of reproductive effects, not for estimation of short-term 50% fetocidal or teratogenic doses, or of threshold levels for continuous intake. Even if enough monkeys could be used to establish such values, there is no assurance that these numbers would be applicable to human populations, in spite of the close biological relatedness of human and nonhuman primates. The few data now available indicate that rhesus macaques are more sensitive (for nonreproductive effects) than any animal tested except the guinea pig. However, although 2,000 people are known to have been occupationally or accidentally exposed to TCDD, and many of these have had chloracne, not one death has been unequivocably ascribed to the toxin. The actual human doses were never known, and the route of entry was more likely the skin than the mouth. Perhaps rhesus macaques are more susceptible than people, but quite likely such a difference is not demonstrable; prudence suggests that the use of experimental findings for human risk assessment be based whenever possible on simian data, meager though they presently are.

QUESTIONS FOR FUTURE RESEARCH

Following is a list of questions about the reproductive effects of TCDD in monkeys for which we need answers. Acquiring these answers will be severely hampered unless the capacity of reference laboratories for low-level analysis for TCDD in animal tissue is expanded. The chemical and mass spectrometric procedures are too complicated and expensive for most biological laboratories, and reference laboratories are presently almost fully occupied with environmental monitoring.

1. What are the effects, in detail, of TCDD on the endocrine events of the menstrual cycle? The reported observation that continuous low-level intake of TCDD by female rhesus macaques depresses circulating estrogen and progesterone levels and causes anovulatory cycles is provocative, but the data are too few to be conclusive.

2. What are the effects of TCDD on spermatogenesis, on the structure and motility of spermatozoa, and on the capacity of sperm to fertilize? The currently emerging technology of _in vitro_ fertilization of primate ova can be useful in answering these questions.

3. Does TCDD pass into the placenta or fetus in early pregnancy, and does enzyme induction occur in fetal tissues and placenta, as it reportedly does in the rat and mouse (15, 16)? Do fetuses that are aborted after maternal exposure to TCDD have developmental aberrations? If gross malformations are not detectable, are there microscopic alterations in fetal tissues, especially those known to be targets for TCDD in rhesus macaques, such as sebaceous glands, thymus, and stomach? Are there chromosomal abnormalities? To answer these questions, one must obtain fetuses surgically before death and expulsion; specimens satisfactory for analysis are almost never recovered after spontaneous abortion.

4. Is TCDD excreted in the milk of lactating mothers? Is TCDD that has been stored in maternal tissue since exposure long before pregnancy mobilized and excreted in the milk in quantities toxic for the nursing infant? I have anecdotal evidence that such storage and later excretion in the milk may occur.

There is ample reason to suspect that current or past exposure to TCDD, at levels eliciting no overt signs of toxicity in the mother, adversely affects fetal development, and the maternal ability to conceive and support a fetus and to nurse an infant. The investigations I recommend with rhesus macaques must necessarily involve intake of TCDD greater than that expected from environmental exposure, for the reasons given above. Answers to these questions, however, may make it possible to estimate a ratio between susceptibility to TCDD's general effects and susceptibility to TCDD's reproductive effects that can be taken into account in risk assessment and regulation.

ACKNOWLEDGMENTS

This work, publication No. 1313 from the Oregon Regional Primate Research Center, was supported by National Institutes of Health grants DE-04222, RR-00163, and RR-05694.

REFERENCES

1. K.D. Courtney and J.A. Moore, Toxicology and Applied Pharmacology 20, 396 (1971).

2. D. Neubert and I. Dillmann, Naunyn-Schmiedeberg's Archives of Pharmacology 272, 243 (1972).

3. F.A. Smith, B.A. Schwetz, K.D. Nitschke, Toxicology and Applied Pharmacology 38, 527 (1976).

4. G.L. Sparschu, F.L. Dunn, V.K. Rowe, Food and Cosmetics Toxicology 9, 405 (1971).

5. E. Giavini, M. Prati, C. Vismara, Environmental Research 27, 74 (1982).

6. W.P. McNulty, American Journal of Primatology 6, No. 1, 41 (1984).

7. M.O. Cheung, E.F. Gilbert, R.E. Peterson, Toxicology and Applied Pharmacology 61, 197 (1981).

8. T. Helder, Toxicology 19, 101 (1981).

9. F.J. Murray, F.A. Smith, K.D. Nitschke, C.G. Humiston, R.J. Kociba, B.A. Schwetz, Toxicology and Applied Pharmacology 50, 241 (1979).

10. D.A. Barsotti, L.J. Abrahamson, J.R. Allen, Bulletin of Environmental Contamination and Toxicology 21, 463 (1979).

11. S.L. Schantz, D.A. Barsotti, J.R. Allen, Toxicology and Applied Pharmacology 48, A180 (1979).

12. R.D. Kimbrough, C.C. Carter, J.A. Liddle, R.E. Cline, P.E. Phillips, Archives of Environmental Health 32, 77 (1977).

13. A.L. Goodman and G.D. Hodgen, Journal of Clinical Endocrinology and Metabolism 49, 469 (1979).

14. C.W. Kocher, N.H. Mahle, R.A. Hummel, L.A. Shadoff, Bulletin of Environmental Contamination and Toxicology 19, 229 (1978).

15. D.L. Berry, T.J. Slaga, N.M. Wilson, P.K. Zachariah, M.J. Namkung, W.M. Bracken, M.R. Juchau, Biochemical Pharmacology 26, 1383 (1977).

16. H. Nau and R. Bass, Toxicology 20, 299 (1981).

REPRODUCTIVE EFFECTS OF THE DIOXINS

Maureen C. Hatch, Ph.D.

Division of Epidemiology
Columbia University School of Public Health
New York, New York 10032

From Public Health Risks of the Dioxins, proceedings of a symposium held on October 19-20, 1983 at The Rockefeller University, New York City. Edited by William W. Lowrance. Copyright, The Rockefeller University, 1984. Published by William Kaufmann, Los Altos, California.

The issue of reproductive hazard has been central to the debate about health risks associated with exposure to chlorinated dioxins. It is a concern voiced by residents of areas where herbicides are used commercially, as well as by Vietnam War veterans exposed through military uses of defoliants. What can scientists tell the anxious public about the effects of such exposure on reproductive function in man? Is the evidence reassuring or does it bear out the fears and claims of those exposed? If uncertainties remain, are the existing data adequate to guide us in deciding directions for future research on populations at risk?

A REVIEW OF THE EPIDEMIOLOGIC EVIDENCE

Besides the experimental work discussed by Drs. Mattison and McNulty, there have been approximately 15 published reports dealing with human exposure to dioxins and various reproductive outcomes. Additional data (as yet unpublished) were recently presented by Vietnamese scientists at a symposium in Ho Chi Minh City in January, 1983. The amount of research done is important since epidemiology depends on a weight-of-evidence approach to overcome the inherent imprecisions of observational research. However, these studies vary greatly in the nature and type of exposure; in the presumed route of exposure (mother, father); in the timing of exposure (prior to conception, during pregnancy); and in the reproductive endpoints considered. This variation in the parameters studied complicates attempts to interpret and generalize the results.

In Table 1, the published studies are grouped by type of exposure (occupational, environmental, industrial accident, military); conditions such as dose and route of administration should be reasonably similar within categories and it may be that the exposure pattern influences the outcome. For each study the research design and the sources of data are described to help weigh the findings. I have also tried, where possible, to specify the biologic model, or presumed model, in order to distinguish crucial similarities and differences in the type, timing and target of exposure, and the range of effects studied.

Table 1. Studies of reproductive effects of human exposure to phenoxy herbicides/chlorinated dioxins.

Exposure/Reference	Study Design/Endpoint(s)	Effects Noted
Occupational: male farmers, forest workers, herbicide applicators, 2,4-D exposure, 1978-1981. (SRI International, 1981)	Frequency of paternal exposure in the two months bracketing conception compared in cases (spontaneous abortions) and controls (exposure) and wife (outcome). Participation rate = 88%. Only events validated in medical records were analyzed. Endpoint: miscarriage.	A borderline association in a subgroup of young forest workers.
Occupational: male chemical workers, dioxin exposure, 1939-1975. (Townsend et al., 1982)	Comparison of outcomes in exposed pregnancies (paternal exposure of a month or more any time prior to conception) and pregnancies to wives of male employees without dioxin exposure. Obstetric data obtained in interview with workers' wives; exposure data from company records. Response rate = 63% for study group, 62% for controls. Endpoints: miscarriage; stillbirth; infant death; birth defects; childhood morbidity.	No association except in one subgroup where an increase in spontaneous abortion was seen. (Definition of exposure assumes damage is irreversible.)
Occupational: male spray applicators, 2,4,5-T exposure, 1969-1980. (Smith et al., 1982)	Comparison of outcomes in exposed pregnancies (paternal exposure in the two years bracketing conception) and conceptions to wives of unexposed agricultural contractors. Data by mail questionnaire. Response rate = 89% for study group, 83% for controls. Endpoints: miscarriage; stillbirth; birth defects.	No association.

Table 1. Continued.

Exposure/Reference	Study Design/Endpoint(s)	Effects Noted
Environmental: forest use of 2,4,5-T in Oregon, 1972-1977. (U.S. EPA, 1979)	Comparison of rates in exposed and unexposed areas. Correlation of monthly herbicide use with monthly miscarriage ratio. Hospital records provided miscarriage data and maternal residence. Company records provided data on 2,4,5-T in a portion of the study area. Data linked by zip code. Endpoint: (hospitalized) miscarriage.	Miscarriage rate higher in exposed area. Miscarriages correlated with spraying pattern (extrapolated from spray data on a portion of the study area).
Environmental: agricultural use of 2,4,5-T in Arkansas, 1948-1974. (Nelson et al., 1979)	Comparison of sex- and race-specific rates over time in areas defined as high, medium or low exposed, based on the extent of rice acreage. Data on outcome from case register of Crippled Children's Service. Endpoint: cleft lip ± palate.	Upward trend over time in both the high and low exposure areas. No significant difference in rates by level of exposure. (Exposure definition ignores use of 2,4,5-T in forested areas.)
Environmental: areas exposed to 2,4,5-T, Australia 1965-1976. (Field & Kerr, 1979)	Temporal and seasonal trends in neural tube rates in New South Wales explored in relation to 2,4,5-T use in all Australia. Data on neural tube defects from a special study. Endpoint: neural tube defects.	Linear correlation between consumption of 2,4,5-T and rates of neural tube defects. Seasonal trends also suggest an association.

Table 1. Continued.

Exposure/Reference	Study Design/Endpoint(s)	Effects Noted
Environmental: areas exposed to 2,4,5-T, Hungary 1969-1976. (Thomas, 1980)	Time trends in selected defects compared with trends in use of 2,4,5-T. Data from birth defects register. Endpoints: neural tube defects; cleft lip; cleft palate; cystic kidney disease.	No linear correlation between neural tube defects and use of 2,4,5-T in Hungary. No increases observed in the other defects studied, despite rising use of 2,4,5-T.
Environmental: areas exposed to 2,4,5-T, New Zealand, 1960-1977. (Hanify et al., 1981)	Comparison of malformation rates in an unexposed period and a period with exposure. Defect rates were plotted against annual average exposure concentration. Hospital records; company records for spray data. Endpoint: birth defects.	An association was observed between exposure and talipes.
Environmental: areas exposed to 2,4,5-T, Hungary, 1970-1980. (Thomas & Czeizel, 1982)	Investigation of time trends and urban/rural patterns in national rates of miscarriage and birth defects in relation to herbicide consumption. Endpoints: miscarriage; birth defects.	Analyses suggest no association.

Table 1. Continued.

Exposure/Reference	Study Design/Endpoint(s)	Effects Noted
Accidental: release of dioxin, Seveso, 1976. (Regianni, 1979)	Comparison of rates in areas zoned by average level of contamination: A=high; B=low; R=no or very low exposure. Rates based on physician notifications. Endpoints: fetal loss; birth defects; infant growth and development; embryomorpholcgy and cytogenetics.	Rates of fetal loss and birth defects increase in late 1976 and 1977 in all zones. In 1977 increase is highest in unexposed area. Infant growth normal; no excess in infectious disease. In 30 induced and 4 spontaneous abortions studied, normal cytogenetics and no clear signs of abnormal development.
Accidental: release of dioxin, Seveso, 1976. (Bisanti et al., 1980)	Comparison of rates in areas zoned by level of contamination (as above). Hospital records provided additional data on miscarriage. Endpoints: miscarriage; birth defects.	Decrease in number of births, especially in Zone A. Increases in miscarriage and birth defects, but rates do not correlate with exposure level.
Military: Agent Orange exposure South Vietnam, 1960-1969. (Cutting et al., 1970)	Comparison of rates in time periods defined by exposure: light spray, 1960-65; heavy spray, 1966-69. Obstetric data from records of selected hospitals. Endpoints: stillbirth; molar pregnancy; congenital defects.	In the Coastal Plain and Delta areas, rates rose in the heavy-spray period. Elsewhere rates were constant or declined.

Table 1. Continued.

Exposure/Reference	Study Design/Endpoint(s)	Effects Noted
Military: Agent Orange, South Vietnam, exposure during pregnancy, 1962-1973. (Kunstadter, 1982)	Consideration of incidence and distribution of defects in time and place, in relation to exposure. Comparison of first-trimester exposure in mothers of affected infants and in mothers of normal infants. Obstetric data from hospital records and personal interviews. Exposure data from HERBS Tapes. Endpoints: all birth defects; cleft palate; perinatal mortality.	Overall decrease in birth defects in the period of exposure. No association of first-trimester exposure with defects. Cleft lip increased relative to other defects in the heavy spraying period. Stillbirth rate increased after exposure ceased.
Military: Service in Vietnam during the period 1962-1972 (Australian troops). (Walsh et al., 1983)	Frequency of military service in Vietnam compared in fathers of cases (children with defects) and controls (normal births) matched for maternal age and time of birth: 8500 case-control pairs. Endpoint: birth defects.	No association between Vietnam service and defects in offspring. (Herbicide exposure was not measured.)

In addition to the studies listed in the table, the major findings by Vietnamese scientists will be considered in summarizing the current state of knowledge of human reproductive effects and in suggesting fruitful areas for future research.

OCCUPATIONAL EXPOSURE

Three studies searched for reproductive impairment in male workers exposed to dioxins in the manufacturing setting or in field applications. In one (Townsend et al., 1982), the exposure is to the presumptive hazard TCDD (2,3,7,8-tetrachlorodibenzo-p-dioxin) as it occurs in the production of chlorophenols. Another (Smith et al., 1982) focuses on men who spray the herbicide 2,4,5-T (2,4,5-trichlorophenoxyacetic acid), which has a TCDD contaminant. In the third study (SRI International, 1981), the exposure under consideration is 2,4-D (2,4-dichlorophenoxyacetic acid), which contains several dioxins but apparently not the 2,3,7,8 isomer; inclusion of this study in the body of evidence on dioxin (insofar as the term is taken to imply 2,3,7,8-TCDD) is thus debatable, at least until the toxicology of the other dioxins is better understood.

All three studies of male work exposure report negative findings, although in two (SRI, 1981; Townsend et al., 1982) significant associations with spontaneous abortion are seen in subgroups of the population. Among men exposed to 2,4-D in various work settings, an adverse effect was observed only in young forest workers; among the chemical production workers exposed to TCDD, the increase in fetal loss was confined to a subgroup married to women with a low risk profile. These could be merely fortuitous findings or the subgroups could represent men whose exposure was at a higher dose or at a crucial period in relation to the pregnancy, but these points are not discussed by the authors. These studies of male-mediated effects, like many others, fail to provide a biologic rationale for the way exposure is defined (e.g., ever exposed, exposed at conception) or an estimate of dioxin levels to which their subjects were exposed, although either or both of these omissions could confound the study results.

There are other weaknesses. For example, in the Townsend study, the response rates, while similar in the exposed and unexposed, are so low (<65%) as to create concern about possible selection bias. Also the control group from the same chemical plant is not well described and it may have been exposed to other adverse agents (although the rates for the various outcomes do not seem unduly high, except perhaps for birth defects at 4.9%). The study of New Zealand spray applicators collected a lengthy reproductive history using a mail questionnaire; this approach is less likely to produce valid and reliable obstetric data than interviews with the workers' wives. Also, the agricultural contractors used as a comparison group in this study are not well described. Neither of these two studies had a sufficiently large sample to rule out with certainty a modest increase in birth defects, the least common of the reproductive endpoints examined.

Three other reports provide indirect evidence concerning parental exposure in the occupational setting and risk of congenital malformation in the offspring. Balarajan and McDowall (1983) used data from Great Britain's birth defects surveillance program to consider paternal occupations with presumptive exposure to dioxin (agriculture, gardening and forestry) in relation to selected defects (spina bifida, anencephaly, orofacial clefts); they report an excess of children with facial clefts for fathers in all three occupations and, among children of gardeners, an increase in neural tube defects as well. Golding and Sladden (1983) used birth certificate data from Oxford and the surrounding area and were unable to confirm the earlier finding in this smaller sample. A study published in Hungary (Szabuka et al., 1980) reports no increase in birth defects among the offspring of female agricultural workers; exposure to 2,4,5-T is assumed by the investigators but, again, is not documented.

ENVIRONMENTAL EXPOSURE

This group of studies considered possible effects on the general population inadvertently exposed to dioxin through herbicide applications on crops or forests. All six studies relied on indirect evidence, either linking available records of herbicide applications with records of reproductive outcomes, or comparing trends over time in herbicide use and reproductive events.

Studies of this kind use data which are easily available and provide a quick, preliminary test of a hypothesis. However, they invariably rest on untested assumptions and are susceptible to problems of misclassification and to the effects of confounding variables. Thus, results may overestimate, underestimate, or even reverse the true relationship between exposure and outcome at the individual level (Robinson, 1950). The study in Arkansas, for instance (Nelson et al., 1979), assumed that forested areas had low or no exposure compared with areas planted in rice, although 2,4,5-T is often sprayed on forests at higher doses than those used on crops (Hatch and Kline, 1981). The results of the study in Oregon (EPA, 1979), which used hospital discharge records to explore the reproductive effects of 2,4,5-T, were confounded by area differences in physician practices for hospitalizing miscarriage patients.

The findings to date on effects of environmental exposure to dioxin, while not consistent, have been largely negative. However, in view of flaws in study design and methods, and the insensitivity of local investigations of low-dose, diffuse exposure, these negative findings are not a sufficient basis for complacency.

ACCIDENTAL EXPOSURE

The data from Seveso, Italy, following the accidental release of dioxin into the environment (Regianni, 1979; Bisanti et al., 1980; Bisanti, 1982) also present problems in interpretation.

After the explosion at the ICMESA plant, zones of exposure were defined based on TCDD levels in the soil. (Zone A, maximally exposed, has a population of approximately 700; Zone B, with moderate exposure, includes about 5000 individuals; and Zone R, with little or no exposure, has a population of almost 32,000.) A system for registering all births was established in the study region, with standardized protocols for diagnosing congenital anomalies. However, there are no good baselines for the period prior to the accident against which comparisons might be made. This makes it difficult to distinguish the effects of improved reporting from real increases in the rates after exposure.

Reliance on comparisons between areas with different exposure levels raises another problem. The validity of area comparisons depends on accurate measures of exposure from zone to zone. However, levels have fluctuated over time and within areas, and despite efforts to prevent it there has been movement of the population in and out of the various zones.

The search for a dose-response pattern has also been thwarted by the sparsity of reproductive events from the area of highest exposure. The birth rate is down in Zone A. Whether this reflects increased contraceptive measures or reduced fertility is not known, but the number of conceptions available for study is indeed small. While increases have been reported in fetal loss and birth defects (particularly spina bifida and multiple malformations), the significance of the observed trends may remain uncertain indefinitely due to the limited number of exposed individuals and the current low birth rate. Even if the trends reflect true effects, it is not possible to know from the Seveso studies whether this is due to maternal exposure, paternal exposure, or both.

MILITARY EXPOSURE

Two studies by U.S. investigators consider the effects of military herbicide use on pregnancy outcome in exposed areas of Vietnam. One (Cutting et al., 1970) compares trends in adverse events in periods of light and heavy exposure. The other (Kunstadter, 1982) assigns exposure status on the basis of the mother's residence at the time of pregnancy; birth defects and perinatal mortality are the outcomes considered. This study finds scant support for the hypothesis that maternal exposure to herbicides during pregnancy is fetotoxic or teratogenic. The Cutting study noted increases in the rate of stillbirth, molar pregnancy, and congenital defects in the Coastal Plain and Delta areas, but not in the other areas studied. Both studies relied on local hospital records which are acknowledged to be incomplete.

Recently, the Australian government published its investigation of birth defects in relation to father's military service in Vietnam (Walsh et al., 1983). No index of exposure of Australian troops to herbicides could be developed; in fact the investigators state that "exposure to herbicides was infrequent and probably very low in Australian troops in Vietnam." The study of 8500 children with congenital defects and 8500 control deliveries, matched for maternal age and time of birth, found no association between father's Vietnam service and malformations in the offspring. Since herbicide exposure for these troops was purportedly infrequent in Vietnam, care should be taken not to construe the finding from this study as evidence for a null effect of dioxin. A similar study is being conducted by the Centers for Disease Control among residents of the Atlanta area and should be reported this spring; this study is able to take some account of putative exposure to Agent Orange.

UNPUBLISHED VIETNAMESE STUDIES

Two types of studies have been carried out by Vietnamese investigators to assess the reproductive effects of herbicide exposure in soldiers and civilians:

1. Studies of veterans' families in the unsprayed North compared outcomes among unexposed women, whose husbands served in the South and were potentially exposed, with those among women whose husbands served only in the North. These studies report an association between presumptive paternal exposure and birth defects in the offspring.

2. Studies of couples in South Vietnam compared reproductive outcomes of those living in sprayed areas with those living in unsprayed areas. Of particular interest here is a striking increase in molar pregnancy among herbicide-exposed women.

A detailed description of each study, including original data tables, is available in the forthcoming proceedings of a symposium on Herbicides in War (Constable and Hatch, in press).

The effects of exposure in the male were examined in three reports. The first, by Drs. Ton Duc Lang, Ton That Tung and Do Duc Van from Viet Duc Hospital in Hanoi, consists of a reproductive survey carried out at agricultural cooperatives set up for veterans. Congenital defects occurred in 71 of 3147 pregnancies to presumptively exposed veterans (2.25%) and in 10 of 2172 pregnancies to unexposed veterans (0.46%), a fivefold difference. The incidence of defects is highest in the first pregnancy following the war (20.4% vs. 5.4%) but persists into subsequent pregnancies. The data are not adjusted for potentially confounding variables such as maternal age, but this would not be influential in the majority of malformations. Although the rate of birth defects in the unexposed is very low, it is compatible with estimates for neighboring countries derived from a WHO study of the same time period.

The second study, by Dr. Nguyen Can and co-workers from the Hanoi Institute for the Protection of Mothers and Newborns, surveys some 40,000 veterans in three areas of North Vietnam. Again the comparisons were between the reproductive outcomes of wives of veterans sent to the South and of veterans remaining in the North, although in this case only pregnancies conceived during the war were considered. Women with drug exposure during pregnancy or a history of tuberculosis, syphilis, or malaria were excluded from the study, and reported outcomes were validated wherever possible. The malformation rate in wartime pregnancies of wives of exposed veterans was 0.64% compared with 0.46% in the pregnancies of wives of unexposed veterans, a small increase of borderline statistical significance ($p=.10$, two-tailed). In both studies facial clefts and anencephaly were more frequent in the children of exposed fathers. However, despite its much larger sample size, the association is not as strong or as statistically significant as in the earlier study.

Dr. Can and colleagues subsequently conducted a case-control study of a sample of surviving children with anomalies, selected at random from the original study population and matched by age to normal offspring of women of the same age, village, and gravidity. The children's families were interviewed about where the father served during the war. This study found a highly significant association between presumptive paternal herbicide exposure (service in the South) and defects in the offspring: an excess risk of 3.5. One possible explanation for the disparity in results between the case-control study and the original survey is that in the earlier study maternal age, gravidity, or some other unevaluated but correlated characteristic may have suppressed the association.

Among studies conducted in the south of Vietnam, which report a range of adverse reproductive outcomes, the most striking finding comes from a 1982 study of molar pregnancy by Drs. Le Thi Diem Huong and Nguyen Thi Ngoc Phuong of Tu Du Gynecological Hospital in Ho Chi Minh City. The frequency of prior herbicide exposure was compared among 100 women hospitalized with molar pregnancies (cases) and 284 women with normal deliveries (controls). Control mothers were matched to cases on maternal age, parity, and social conditions. Almost 57% of cases reported exposure to herbicide spraying as compared with 9.8% of controls, an excess risk of more than tenfold.

Routine hospital statistics since the war do not show any increase in molar pregnancy until 1976, about 10 years after the heavy spraying. The average age of the study subjects during intense herbicide exposure was 13-18. This suggests either a delayed effect of acute, wartime exposure or an effect of chronic exposure to dioxin residues. Since the husbands of these women are likely to share their exposure, do we know whether the effect is maternally or paternally mediated? The geographic trends in molar pregnancy (increases in the South but not in the North) seem to indicate that the adverse effect is due to exposure of the mother.

Molar pregnancies, generally much more common among Asian than Caucasian women, are conceptions without an embryo, but with grossly swollen placental villi. They are androgenetic in origin, arising from fertilization by a normal sperm of an ovum in which the nucleus is either absent or inactivated (Jacobs et al., 1980). Molar pregnancy is associated with some form of malignant sequelae (persistent mole, invasive mole, or choriocarcinoma) in about 15% of cases (Goldstein and Berkowitz, 1982). Determination of dioxin levels either in the mole or subsequent choriocarcinoma or in the fat or other tissues of the patient might confirm the possible association between herbicides and hydatidiform mole/choriocarcinoma, and also perhaps clarify the timing of exposure that is crucial in producing the underlying germ cell damage.

LEADS AND LESSONS: PROSPECTS FOR THE FUTURE

Studies of Vietnamese civilians and soldiers and, to a much lesser extent, Seveso residents suggest an association between exposure to dioxin and impaired reproductive function in man. Effects noted include spontaneous abortion, neural tube defects, oral clefts, limb anomalies, multiple malformations, molar pregnancy, and perhaps infertility. On the other hand, studies of exposed workers and residents of forest or farm areas suggest that, overall, dioxin is not hazardous to reproduction. Which studies approximate the true risk? There have been weaknesses in research design and methods in studies reporting negative results as well as in studies reporting positive results; however, without access to the protocols and raw data, it is difficult to measure the extent of bias that has been introduced. What then can we conclude? Have we learned anything to guide future reproductive research, either in the laboratory or in studies of Vietnam veterans and residents of dioxin-contaminated areas such as Times Beach, Missouri?

Despite the considerable uncertainties, some leads appear worth pursuing, and there are ways to improve study design that may help move the research forward.

1. In terms of <u>research design</u>, the studies reviewed here suffer from a lack of precision in the measurement of exposure and a lack of clarity as to the biologic model being tested. There is much to be gained from increasing the level of specificity, both in relation to exposure and to biologic effect.

Wherever possible, a dose index should be developed; at the very least, a mean or range of values for dose level of dioxin and associated chemicals should be a datum in the research report. Since biologic response may vary by route of absorption, an attempt should be made to differentiate inhalation exposure and skin contact from ingestion. The resulting refinement in exposure data should deepen our understanding of the chlorinated dioxins.

Equally important is the need to be explicit and systematic in exploring the pathophysiologic mechanism. Plausible biologic models for adverse effects should be defined and tested. This is especially important in studies of male exposure and birth defects, where the association (should it exist) may be confined to exposure at certain times in relation to spermatogenesis or pregnancy, as well as to certain categories of congenital malformation. The design of future studies of paternally-mediated effects should accommodate these possibilities.

In a similar vein, one interpretation of the molar pregnancy data from Vietnam suggests that the phenomenon can result from maternal exposure considerably prior to conception. Although it is not clear why oocytes should be more vulnerable then, the issue of timing should be addressed in subsequent studies of this association.

2. Several interesting leads provided by the earlier research might be followed up in laboratory and epidemiologic studies:

Assuming an animal model is available, an attempt might be made to replicate the molar pregnancy finding, or at least to explore the effects of exposure on female oocytes. Additional epidemiologic studies could be done, not only in Vietnam but also in the U.S., where there are concentrations of Vietnamese refugees or of American women with phenoxy herbicide exposure.

The possibility that dioxin interferes with fertility is raised by the experimental work reviewed in this symposium and indirectly by data from Seveso. This area needs to be investigated using a variety of indices, such as reduced birth rate, conception delay, problems in menarche, and semen and hormone analyses. Such studies could contribute considerably to understanding the possible mechanisms underlying other reproductive effects that have been observed.

ADDENDUM

Since writing this paper, the U.S. Air Force has released its first report on morbidity in men involved in the herbicide spraying missions in Southeast Asia, known as Operation Ranch Hand (Lathrop et al., 1984). Birth defects were reported more frequently in the offspring of Ranch Handers, significantly so when minor anomalies are included. Neonatal death and physical handicap were also reported more often for children born to Ranch Hand personnel. No differences were observed in fertility or fetal loss between Ranch Handers and cargo-mission personnel who were not occupationally exposed to Agent Orange.

ACKNOWLEDGMENTS

I would like to thank Dr. John Constable, with whom many of these points were developed and debated, and Dr. Zena Stein, who read the manuscript carefully and asked many clarifying questions.

REFERENCES

Balarajan, R. and McDowall, M. Congenital malformations and agricultural workers (letter). Lancet i:1112-13, 1983.

Bisanti, L., Bonetti, F., Caramaschi, F., Del Corno, G., Fara, G.M., Favaretti, C., Giambelluca, S.E., Marni, E., Montesarchio, E., Puccinelli, V., Remotti, G., Volpato, C. and Zambrelli, E. Experiences from the accident of Seveso. Acta Morphologica Academiae Scientiarum Hungaricae 28:139-57, 1980.

Bisanti, L. Le malformazioni congenite. Sapere 848:68-74, 1982.

Constable, J. and Hatch, M. Herbicide exposure and reproductive outcome. In: Herbicides in War: The Long-Term Effects on Man and Nature, A.W., Westing, ed. New York: Taylor & Francis, 1984 (in press).

Cutting, R.T., Phuoc, T.H., Ballo, J.M., Benenson, M.W. and Evans, C.H. Congenital Malformations, Hydatidiform Moles and Stillbirths in the Republic of Vietnam, 1960-1969. U.S. Government Printing Office, Washington DC, 1970.

Field, B. and Kerr, C. Herbicide use and incidence of neural-tube defects (letter). Lancet i:1341-42, 1979.

Golding, J. and Sladden, T. Congenital malformations and agricultural workers (letter). Lancet i:1393, 1983.

Goldstein, D.P. and Berkowitz, R.S. Nonmetastatic and low-risk metastatic gestational trophoblastic neoplasms. Seminars in Oncology 9:191-97, 1982.

Hanify, J.A., Metcalf, P., Nobbs, C.L. and Worsley, K.J. Aerial spraying of 2,4,5-T and human birth malformations: an epidemiological investigation. Science 212:349-51, 1981.

Hatch, M. and Kline, J. Spontaneous Abortion and Exposure During Pregnancy to the Herbicide 2,4,5-T. U.S. Environmental Protection Agency, Washington DC (EPA 560/6-81-006), 1981.

Jacobs, P.A., Wilson, C.M., Sprenkle, J.A., Rosenshein, N.E. and Migeon, B.R. Mechanisms of origin of complete hydatidiform moles. Nature 287:714-16, 1980.

JRB Associates. Review of Literature on Herbicides, including Phenoxy Herbicides and Associated Dioxins, Vol. 1. Veterans Administration, Washington DC, 1980.

Kunstadter, P. A Study of Herbicides and Birth Defects in the Republic of Vietnam: An Analysis of Hospital Records. National Academy of Sciences, National Academy Press, Washington DC, 1982.

Lathrop, G.D., Wolfe, W.H., Albanese, R.A. and Moynahan, P.M. Project Ranch Hand II. An Epidemiologic Investigation of Health Effects in Air Force Personnel Following Exposure to Herbicides: Baseline Morbidity Study Results. San Antonio, Texas: U.S. Air Force School of Aerospace Medicine, Aerospace Medical Division, Brooks Air Force Base, 1984.

Nelson, C.J., Holson, J.F., Green H.G. and Gaylor, D.W. Retrospective study of the relationship between agricultural use of 2,4,5-T and cleft palate occurrence in Arkansas. Teratology 19:377-83, 1979.

Reggiani, G. Estimation of the TCDD toxic potential in the light of the Seveso accident. Archives of Toxicology (Suppl):291-302, 1979.

Rehder, H., Sanchioni, L., Cefis, F. and Gropp, A. (Pathological and embryological studies on abortion cases related to the Seveso accident.) Schweizerische Medizinische Wochenschrift 108:1617-25, 1978.

Robinson, W.S. Ecological correlations and the behavior of individuals. American Sociological Review 15:351-57, 1950.

Smith, A.H., Fisher, D.O., Pearce, N. and Chapman, C.J. Congenital defects and miscarriages among New Zealand 2,4,5-T sprayers. Archives of Environmental Health 37:197-200, 1982.

SRI International. A Case-control Study of the Relationship between Exposure to 2,4-D and Spontaneous Abortions in Humans. Report for the National Forest Association, Washington DC, August, 1981.

Szabuka, M. et al. (Reproductive examination of the female workers of large agricultural concerns.) Magyar Noorvosok Lapja 43:527-34, 1980.

Thomas, H.F. 2,4,5-T use and congenital malformation rates in Hungary (letter). Lancet ii:214-15, 1980.

Thomas, H.F. and Czeizel, A. Safe as 2,4,5-T? (letter). Nature 295:276, 1982.

Townsend, J.C., Bodner, K.M., Van Peenen, P.F., Olsen, R.D. and Cook, R.R. Survey of reproductive events of wives of employees exposed to chlorinated dioxins. American Journal of Epidemiology 115: 695-713, 1982.

U.S. Environmental Protection Agency. Six Years' Spontaneous Abortion Rates in Oregon Areas in Relation to Forest 2,4,5-T Spray Practices. EPA, Washington DC, 1979.

Walsh, R.J., Donovan, J.W., Adena, M.A., Rose, G. and Battistutta, D. Case-control Study of Congenital Anomalies and Vietnam Service (Birth Defects Study). Canberra: Australian Government Publishing Services, 1983.

IMMUNOLOGICAL EFFECTS FOLLOWING EXPOSURE TO 2,3,7,8-TETRACHLORODIBENZO-*p*-DIOXIN: A REVIEW

Jack H. Dean, Ph.D.
Lloyd D. Lauer, M.S.

Chemical Industry Institute of Toxicology
Research Triangle Park, North Carolina 27709

From *Public Health Risks of the Dioxins*, proceedings of a symposium held on October 19-20, 1983 at The Rockefeller University, New York City. Edited by William W. Lowrance. Published by William Kaufmann, Los Altos, California.

INTRODUCTION

When unperturbed, the immune system provides resistance to infectious agents, and provides homeostasis of leukocyte maturation, immunoglobulin synthesis, and immune surveillance against arising neoplastic cells. The interaction of xenobiotics, including the dibenzodioxins, with immunoresponsive cells may perturb cellular differentiation, homeostasis, or cell/cell cooperation, which may result in undesirable effects expressed as immunosuppression, hypersensitivity, and alterations of host defense mechanisms.

The immune response is a series of complex events which follow the introduction of an antigenically foreign material or infectious agent into an immunocompetent host. There are two major types of immune responses: cell mediated immunity (CMI), and humoral immunity (HI). CMI represents responses involving specifically sensitized, thymus-dependent lymphocytes (T-cells). The induction of CMI proceeds by a sequential differentiation of small lymphocytes into large pyroninophilic cells (activated or sensitized) which will ultimately divide giving rise to cells responsible for immunologic memory and various effector functions. Effector cell functions include cytotoxicity for infectious agents and virus-infected or neoplastically-transformed cells; suppressor activity down-regulating both T- and B-cell responses; helper activity facilitating B-cells in antibody responses; and, the elaboration of lymphokines which are humoral factors mediating many CMI functions (e.g., interferon, interleukins, chemotactic factor, macrophage activation factor).

Humoral immunity (HI) involves the production of specific serum antibodies (immunoglobulins, Ig) by bursa-equivalent lymphocytes termed plasma cells (B-cells) following antigenic stimulation. Serum antibodies interact with the specific eliciting agent (antigen) when it is again encountered and help to eliminate foreign material or infectious agents from the host. In contrast to CMI, HI is responsible for host resistance to a somewhat different spectra of infectious agents, usually those tending to be nonpersistent. Both CMI and HI have been perturbed by exposure of rodents to dibenzodioxins.

Halogenated aromatic compounds represent a family of toxic chemicals often produced as unwanted contaminants in the synthesis of other economically important industrial chemicals (see review 20). These pollutants are a concern to human health because of their high magnitude of toxicity, ubiquitous occurrence, persistence in the environment, and tissue affinity. The most toxic of the halogenated aromatic compounds are the dibenzofurans and dibenzodioxins which occur as contaminants produced during the synthesis of other polyhalogenated aromatics. There is substantial evidence that a number of isomers of this family are carcinogenic, teratogenic, and immunotoxic (see reviews 13,19,20). This chapter will focus on the immunological effects produced by the prototype dibenzodioxin - 2,3,7,8-tetrachlorodibenzo-*p*-dioxin. The reader is referred to several recent reviews for a comprehensive discussion of the literature concerning other members of this class of chemicals which have been found to perturb the immune system (see reviews 5,7,32).

DIBENZODIOXIN TOXICITIES

The dibenzodioxin 2,3,7,8-tetrachlorodibenzo-*p*-dioxin (TCDD) is the most biologically active member of this group and is produced as an unwanted contaminant in the commercial synthesis of 2,4,5-trichlorophenol, a precursor to the herbicide 2,4,5-trichlorophenoxyacetic acid (2,4,5-T), and the bacteriostat hexachlorophene (23). Human exposure to TCDD has resulted from industrial accidents, chemical wastes, and widespread use of defoliants such as Agent White and Agent Orange, an esterified mixture of 2,4-dichlorophenoxyacetic acid and 2,4,5-T (see review 1). Agent Orange contained up to 30 ppm TCDD, and it is estimated that during the course of the military defoliation program in Vietnam (1962-1970), approximately 100 kg of TCDD were released into the environment (10). Large scale accidental exposure of humans to TCDD occurred in Seveso, Italy, a small town north of Milan, where in 1976 a reactor in a plant manufacturing 2,4,5-trichlorophenol overheated, spreading 1 to 1.5 kg of TCDD over a 700 acre area (12); in Nitro, West Virginia, where a similar accident (1949) resulted in exposure of workers involved in the manufacture 2,4,5-T and those involved in the plant clean-up; and possibly among inhabitants of Times Beach, Missouri in recent years.

Studies on the mechanism of action of TCDD in animals have shown that it is the prototype for a large series of isosteric halogenated dibenzo-p-dioxins, dibenzofurans, azo- and azoxybenzenes, and biphenyls. Characteristic toxic responses to TCDD in animals include a slow wasting syndrome, chloracne, teratogenesis, and thymic atrophy (13). The available toxicology data (see review 19) indicate that the lethality associated with acute exposure (oral LD_{50} of less than 1 ug/kg in the guinea pig) as well as the expression of a particular pattern of toxic effects seen with lower doses is species specific (23).

Rats, mice, and guinea pigs exposed to TCDD demonstrate severe thymus atrophy histologically characterized as cortical lymphoid depletion similar to that induced by cortisone (18,32,33). The offspring of TCDD-exposed animals often exhibit a "wasting syndrome" (8,28) much like that seen in neonatally thymectomized rats. Depletion of T-cell dependent areas of the thymus is accompanied by suppression of CMI functions when exposure occurs during ontogenesis of the immune system (15,35). These effects are less severe if exposure occurs during adult life (2). Lymph node atrophy, leukopenia, and depressed bone marrow cellularity have also been frequently reported following pre/postnatal exposure to TCDD in rodents (15). The ED_{50} (dose producing 50% maximal response) for the induction of thymic atrophy in sensitive mouse strains is approximately 10 nmol/kg (22).

EFFECTS OF TCDD OF CELL-MEDIATED IMMUNITY

A common observation in rodents following adult or pre/postnatal exposure to TCDD has been the suppression of cell-mediated immunity (Table 1) generally expressed as a dose-dependent impairment of delayed cutaneous hypersensitivity to sensitizing antigens; proliferative responses to T-cell and B-cell mitogens and allogeneic leukocytes in unidirectional mixed leukocyte cultures (MLC); and, graft-versus-host and allograft rejection responses.

Table 1. Cell-mediated immune alterations following exposure to TCDD.

Strain/Species	Total Dose[a]	Route/Duration of Exposure	Parameter and Results (Reference)
Lymphocyte Blastogenesis:			
F344 Rat	5 ug/kg	Gavage/wk x 4	PHA - decreased (35)
C57BL6	25 ug/kg	Gavage/wk x 4	PHA - decreased (35)
Swiss	5 ppb	Feeding/pre & post-natal	Con A and LPS - no effect (28)
B6C3F1	15 ug/kg	Gavage/pre & post-natal	PHA and Con A - decreased (15)
CD-1	1-10 ug/kg	Gavage/wk x 8 or single	PHA and PWM - decreased (26)
C57BL/6	30 ug/kg	ip/single	Con A and LPS - no effect (30)
B6C3F1	1 ug/kg	ip/single	PHA, Con A, LPS and MLC - decreased (4)
Delayed Type Hypersensitivity:			
Swiss	5 ppb	Feeding/pre & post-natal	DNFB - decreased (28)
C57BL/6	40 ug/kg	ip/wk x 4	Oxazolone - decreased (2)
Guinea pig	.04 ug/kg	Gavage/wk x 8	Tuberculin - decreased (36)
F344 rat	5 ug/kg	Gavage/wk x 6	Tuberculin - no effect (36)
	5 ug/kg	Gavage/pre & post-natal	Oxazolone - decreased (8)
Graft vs Host:			
F344 Rat	1 ug/kg	Gavage/pre & post-natal	No effect (36)
	5 ug/kg	Gavage/wk x 4	Decreased (35)
C57BL/6	30 ug/kg	ip/single	No effect (30)

Table 1. Continued.

Strain/Species	Total Dose[a]	Route/Duration of Exposure	Parameter and Results (Reference)
Graft Rejection:			
C57BL/6	2 ug/kg	Gavage/pre & post-natal	Decreased (35)
F344 Rat	5 ug/kg	Gavage/wk x 4	Decreased (35)
T-Cell Cytotoxicity:			
C57BL/6	4 ng/kg	ip/wk x 4	Decreased (2)
B6C3F1	10 ug/kg	ip/single	No effect (4)
NK Cytotoxicity:			
C57BL/6	30 ug/kg	ip/wk x 4	No effect (17)
B6C3F1	10 ug/kg	ip/single	No effect (4)

[a]Minimal dose producing significant ($p < 0.05$) effect or maximal dose producing no effect.

Tumoricidal effector functions associated with natural killer cells have not been affected in lymphocytes from TCDD-exposed mice following adult exposure. However, the induction of T-cells cytotoxic for tumor target cells was found to be impaired in studies by Clark et al. (2) following a total exposure dose of 4 ng/kg over a four-week period. These dosages were below those levels significantly altering other CMI responses (2). The generation of precursor CTL and production of the proliferation-inducing lymphokine, interleukin 2 (IL2), by helper T-cells were unaltered by TCDD exposure (2). The suppression of CTL effector function is believed due to induction of CTL-specific suppressor T-cells (3). In a preliminary study (4) of adult B6C3F1 mice exposed to a single 10 ug/kg dose of TCDD (2500-fold greater than the dose used in Clark's studies), CTL activity was not suppressed 7-10 days following exposure, although other CMI functions were altered. Since the kinetics of CTL suppression are undefined in Clark's studies (2,3), this important issue of the effect of TCDD on CTL-mediated tumor resistance remains to be resolved. Finally, caution is advised in interpreting immune dysfunction in studies involving prenatal and/or postnatal exposure since some, but not all, of these studies noted a high degree of fetal toxicity.

TCDD apparently does not suppress CMI indirectly through endogenous corticosteroids since thymus atrophy was observed in adrenalectomized or hypophysectomized rats given TCDD (29,32). Likewise, TCDD is not cytotoxic for lymphocytes *in vitro* (16) nor does it appreciably alter macrophage function (Table 2).

Table 2. Macrophage function following exposure to TCDD.

Species	Function	Results (Reference)
Swiss	Number of Peritoneal Macrophages, Phagocytosis, NBT Reduction	No effect (34)
C57BL/6	Number of Peritoneal Macrophages, Elicited Cytotoxicity	Reduced No effect (17)
B6C3Fl	Number of Peritoneal Macrophages and Cytostasis of Tumor Cells	No effect (4)

EFFECTS OF TCDD ON ANTIBODY-MEDIATED IMMUNITY

The effect of TCDD on antibody-mediated immunity has been assessed by quantitating the specific serum antibody titer or the number of IgM plaque forming cells (PFC) produced following challenge with eliciting antigen (Table 3). Both parameters have been markedly depressed following adult or pre/postnatal exposure to TCDD in multiple inbred mouse strains and guinea pigs. In contrast to C57BL/6 or C3H mice, antibody responses were not suppressed following TCDD exposure in DBA/2 mice, a poorly *Ah*-inducible mouse (31). Suppression of antibody-mediated immunity appears to be a more sensitive indicator of TCDD toxicity in adult mice than is suppression of CMI. The magnitude of suppressed antibody immunity in inbred mouse strains, like suppression of T-cell immunity, appears to require the coordinate expression of several inducible enzymes, including cytochromes(s) P_1-450, which are regulated by a single genetic locus, designated the Ah. Mapping with the Ah locus is another gene product, the TCDD cytosolic receptor protein, which has been identified in the thymic cytosol from C57BL/6 but not from DBA/2 mice (6,23,24). The importance of receptor protein expression in TCDD toxicity is further demonstrated by the observation that the ED_{50} value for antibody PFC suppression differs by greater than 30-fold (1 ug/kg > 30 ug/kg) between C57BL/6 and DBA/2 mice. A mutation in the Ah locus of DBA/2 mice has apparently resulted in the expression of a TCDD receptor binding protein with reduced affinity, which probably accounts for the minimal TCDD immunotoxicity observed in this strain.

Table 3. Altered antibody-mediated immunity following exposure to TCDD.

Strain/Species	Total Dose[a]	Route/Duration of Exposure	Antigen	Effect (Reference)
Antibody Forming Cells:				
Swiss	5 ppb	Feeding/pre- and postnatal	SRBC	Decreased (28)
C57BL/6	1.2 ug/kg	ip/single	SRBC	Decreased (30)
		7-24 days	SIII (TI)	Decreased (30)
	4 ng/kg	ip/wk x 4	SRBC	Decreased (2)
C3H	6 ug/kg	ip/single	SRBC	Decreased (31)
DBA/2	30 ug/kg	ip/single	SRBC	No effect (31)
B6C3F1	1 ug/kg	ip/single	SRBC	Decreased (4)
			TNP-LPS (TI)	Decreased (4)
Serum Antibody Titer:				
Guinea pig	0.2 ug/kg	Gavage/wk x 4	Tetanus Toxoid	Decreased (36)
Swiss	10 ppb	Feeding/5 wks	SRBC	Decreased (11)
			Tetanus Toxoid	Decreased (11)
	5ppb	Feeding/pre- and postnatal	SRBC	No effect (28)

Table 3. Continued.

Strain/Species	Total Dose[a]	Route/Duration of Exposure	Antigen	Effect (Reference)
Serum Antibody Titer: (continued)				
F344	5 ug/kg	Gavage/pre- and postnatal	BGG	No effect (8)

[a]Minimal dose producing significant ($p < 0.05$) effect or maximal dose producing no effect. TI=T-cell independent.

EFFECTS OF TCDD ON HOST RESISTANCE

Alterations of host resistance to challenge with infectious agents or transplantable tumor cells has been shown to correlate with immune dysfunction following exposure to xenobiotics (see review 5). Thigpen et al. (27) observed that exposure of C57BL/6 mice to TCDD (1 ug/kg by gavage, weekly x 4) resulted in increased susceptibility to challenge with the gram-negative bacterium *Salmonella*. Subsequent studies have reported variable effects on host resistance to challenge with bacteria, tumor cells, or viruses following TCDD exposure (Table 4). Studies on the effect of TCDD on host resistance are complicated by variables in pre/postnatal versus adult exposure; amount and duration of TCDD exposure; severity and type of immune dysfunction; and mechanism of immune resistance required for the particular pathogen (e.g., T-effector cells, neutralizing antibody, ADCC, etc.). At present, the issue of host resistance effects following TCDD exposure is poorly resolved.

Table 4. Altered host resistance to infectious agents and tumor challenge following exposure to TCDD.

Agent	Total Dose[a]	Route/Duration of Exposure	Strain	Mortality (Reference)
Bacteria:				
Salmonella	1 ug/kg	Gavage/wk x 4	C57BL/6	Increased (27)
	50 ppb	Feeding/ 8 wk	Swiss	Increased (11)
Listeria	50 ug/kg	Gavage/wk x 4	C57BL/6	No effect (34) (organisms per spleen)

Table 4. Continued.

Agent	Total Dose[a]	Route/Duration of Exposure	Strain	Mortality (Reference)
Bacteria: (continued)				
Listeria	5 ug/kg	Gavage/pre- and postnatal	B6C3F1	Increased (15)
	5 ppb	Feeding, pre- and postnatal	Swiss	No effect (28)
	50 ppb	Feeding/8 wk	Swiss	Increased (11)
	10 ug/kg	ip/single	B6C3F1	No effect (4)
Virus:				
Herpes Suis	20 ug/kg	Gavage/wk x 4	C57BL/6	No effect (27)
Tumor Cells:				
PYB6	1 ug/kg	Gavage, pre- and postnatal	B6C3F1	Increased (15)

[a]Minimal dose producing significant ($p < 0.05$) effect or maximal dose producing no effect.

IMMUNE ONTOGENESIS

It has been generally held that exposure to TCDD during immune ontogenesis, which occurs in mice and rats during very late prenatal and early neonatal life, has a more profound effect on immune function than adult exposure (35). This belief is supported by the observation that exposure of guinea pigs to TCDD during the neonatal period was found to suppress delayed hypersensitivity responses, while adult exposure in rats did not alter this parameter. Likewise, thymus atrophy, and to a lesser extent lymph node atrophy, has been more severe and frequently associated with altered T-cell subpopulations and suppressed CMI following prenatal and/or neonatal exposure rather than adult exposure (2,3,35,36). This may suggest that TCDD-induced thymic atrophy results from actions on progenitor lymphoid cells in the bone marrow (32), or through altered diffentiation of intrathymic precursor cells, specifically by a direct action on thymic epithelium (24,32), or both (9). It is becoming apparent that adult exposure can likewise suppress antibody-mediated immunity and, to a lesser extent, CMI function (4,30). This effect may also be mediated at the level of the thymus, since the thymus is now recognized to have a continuing hormonal role in maintaining the immune integrity of the adult animal. In some species, in utero exposure (via maternal dosing) appears to be necessary to induce maximum immunosuppression (8,15,35). At higher dosages, antibody responses and bone marrow stem cell numbers are depressed in most species (15,35). Administration of TCDD in utero also results in decreased resistance of offspring to bacterial and tumor cell challenge correlating with altered CMI (15).

IMMUNE EFFECTS FOLLOWING TCDD EXPOSURE IN ADULT B6C3F1 MICE

TCDD has been regarded as primarily immunotoxic for the developing immune system during the pre/postnatal period (i.e., immune ontogenesis). To further characterize effects of TCDD exposure on adult immune function and host resistance, adult B6C3F1 mice (7-10 weeks old) were exposed to TCDD by a single intraperitoneal injection at doses of 0.1, 1.0, and 10.0 ug/kg of body weight. Seven to 10 days following exposure, lymphoid organ weights, proliferative and cytotoxic responses, antibody plaque forming cell response, and host resistance to bacterial and tumor cell challenges were evaluated. At the 10 ug/kg dose of TCDD, thymus weights were significantly reduced (39%) while spleen and body weights were unchanged (Table 5). Lymphocyte proliferation in response to T- and B-cell mitogens was significantly suppressed up to 50%. The one-way mixed lymphocyte culture (MLC) response was also depressed (40%) at the highest exposure. In contrast, natural killer (NK) cell and cytotoxic T-lymphocyte (CTL) mediated tumor cytolysis were unaltered. The number of IgM antibody plaque forming cells produced in response to sheep erythrocytes (a T-dependent antigen) and TNP-Ficoll (a T-independent antigen) was depressed up to 86% and 49%, respectively.

Table 5. Effect of TCDD on immune function in adult B6C3F1 mice.

Parameter	Control	TCDD Dose (ug/kg) 0.1	1.0	10
Body Weight (g)	22.0 ± 0.7	22.4 ± 0.4	23.3 ± 0.5	22.2 ± 0.3
Spleen Weight (mg)	91.0 ± 6.2	88.5 ± 4.0	103.0 ± 7.7	81.3 ± 5.7
Thymus (mg)	57.5 ± 7.4	48.5 ± 3.7	54.0 ± 5.6	35.3 ± 5.8[a]
IgM PFC (per 10^6 splenocytes)	2364 ± 176	2295 ± 213	1008 ± 343[b]	444 ± 135.3[b]
Proliferative Responses: (cpm ± SEM x 10^{-3})				
PHA	159.0 ± 8.8	152.5 ± 9.6	103.6 ± 11.1[b]	97.0 ± 12.1[b]
Con A	145.3 ± 4.4	166.3 ± 6.6	109.3 ± 9.7[a]	87.0 ± 5.7[b]
LPS	43.7 ± 2.5	40.4 ± 3.5	22.5 ± 3.3[b]	20.5 ± 2.0[b]
MLC	89.7 ± 5.9	84.0 ± 12.3	70.0 ± 9.2	52.7 ± 7.0[b]
Tumoricidal Responses: (% Release)				
NK Cytolysis	15.5	12.5	16.8	13.5
Cytotoxic T-Cell Cytolysis	83	ND	ND	88
Challenge Studies:				
Listeria monocytogenes (% Mortality)	10	0	0	0
PYB6 Tumor Cells (% Tumor incidence)	35	10	10	50
B16F10 Melanoma (Lung nodules ± SEM)	48.3 ± 17.6	66.4 ± 11.5	91.0 ± 17.2	87.7 ± 20.8
CPM ^{125}IUDR Incorporation	978 ± 460	500 ± 111	1031 ± 401	736 ± 160

[a]Significantly different from controls $p \leq 0.05$.
[b]Significantly different from controls $p \leq 0.01$.

Resistance to challenge with the bacterium Listeria monocytogenes, PYB6 sarcoma cells, or B16F10 melanoma cells was unaltered by TCDD exposure in adult B6C3F1 mice (Table 5). Previous studies have demonstrated that in utero or perinatal exposure of mice to TCDD produces severe thymus atrophy and immune dysfunction (15). The current study indicates that adult exposure to TCDD produces similar thymus atrophy, decreases in lymphocyte proliferation, and suppression of antibody plaque forming cell numbers, as does pre/postnatal exposure, without affecting in vitro tumor cell cytolysis (i.e., NK and CTL) and host resistance to tumor and bacterial challenges. Studies are currently in progress utilizing immunorestorative agents to further characterize the immune dysfunction produced by TCDD in adult mice.

EFFECT OF TCDD ON THE IMMUNE STATUS OF HUMANS

In a report characterizing the immune status of 44 school children residing in the TCDD-contaminated area of Seveso, Italy, it was revealed that 20 children exhibited chloracne (a classic sign of TCDD toxicity) (25). No immune alterations were noted in serum immunoglobulin levels and circulating complement levels. The lymphoproliferative responses to T- and B-cell mitogens were not depressed, but were significantly enhanced in children exhibiting chloracne. Enhanced proliferative responses may be artifactual because these studies were performed in multiple laboratories, or they may mimic the enhanced proliferative responses to mitogens seen at low TCDD exposures in rodent studies (15). In an earlier evaluation of British workers in a chemical manufacturing plant who were accidentally exposed to TCDD, serum IgD and IgA levels were reduced as were lymphocyte responses to T-lymphocyte mitogens (Ward, unpublished report). A correlation was suggested between chloracne and altered immune status, although these data have never been subjected to careful peer review or published. Bekesi and his associates at the Mt. Sinai School of Medicine have recently evaluated workers involved in the cleanup of a TCDD accident in 1949 at a chemical plant in Nitro, West Virginia; these data remain too preliminary for interpretation at present. The Air Force has just reported on a health study of a well-defined cohort of Project Ranch Hand II personnel who were exposed to herbicide during the course of aerial herbicide dissemination missions (Ranch Hand) in Vietnam (14). Of this cohort, 297 "Ranch Handers" were randomly selected for immune function evaluation along with 295 matched controls. In this study, T- and B-lymphocyte subsets were enumerated and the proliferative response of lymphocytes to mitogen and antigen stimuli was assessed. Although this immunology study has some technical flaws, no gross immunological alterations were noted between the herbicide-exposed and comparison groups.

Finally, in yet another unpublished report (21), 47 men exposed from 1-90 days to a crude batch of chlorophenols contaminated with dibenzodioxin and exhibiting chloracne were evaluated for immune function. This group was also reported to have increased proliferative responses to T- and B-cell mitogens with an accompanying altered ratio of T- and B-cells (21). Neutrophil function was believed to be suppressed, although this report is preliminary, unpublished, and has not been subjected to peer review. At present, all published reports on immune status of humans exposed to TCDD have reported no gross immunological abnormalities, although most unpublished data have suggested some immune alterations. Additional studies of cohorts with well-documented TCDD exposure will be required before this issue can be resolved, although at present no indication of serious immune dysfunction or host resistance impairment has been observed in humans inadvertently exposed to TCDD.

ABSTRACT

The dibenzodioxin 2,3,7,8-tetrachlorodibenzo-p-dioxin (TCDD) is produced as an unwanted contaminant in the commercial synthesis of 2,4,5-trichlorophenol, a precursor to the herbicide 2,4,5-trichlorophenoxyacetic acid, and the bacteriostat hexachlorophene. Exposure of rodents to TCDD as adults or perinatally during immune ontogenesis results in thymus atrophy and suppression of bone marrow, cell-mediated immunity (CMI), humoral immunity (HI) and, with variable results, host resistance to infectious agents or transplantable tumor cells. Alterations of thymus development, CMI, and host resistance are more profound when exposure occurs during the prenatal and/or postnatal periods when maturation of thymus-dependent lymphocytes is maximal. CMI alterations observed include depressed delayed cutaneous hypersensitivity reactions; lymphoproliferative responses to T- and B-cell mitogens, graft rejection and, with variable results, induction of cytoxic T-lymphocytes. Natural killer cell tumoricidal function is unaltered by TCDD exposure. Humoral immunity, expressed as the production of serum antibody or antibody producing cells following antigen challenge, has been profoundly suppressed during both adult and pre/postnatal TCDD exposure. In some studies, challenge of TCDD-exposed animals with infectious agents or tumor cells has resulted in increased mortality depending on the duration of TCDD exposure, challenge agent selected, and other variables. The available data suggest that TCDD-induced thymus atrophy and immune alterations may result from direct actions on peripheral lymphocytes or progenitor lymphoid cells in the bone marrow, and through altered differentiation of intrathymic precursor cells, specifically by a direct action on thymic epithelium. Finally, in the limited number of published studies of immune function in humans accidentally exposed to TCDD, significant immune impairment has not been observed.

REFERENCES

1. Dioxins. Chemical and Engineering News 61(23): 1-84 (1983).

2. Clark, D.A., Gauldie, J., Szewczuk, M.R., and Sweeney, G. Enhanced suppressor cell activity as a mechanism of immunosuppression by 2,3,7,8-tetrachlorodibenzo-p-dioxin. Proceedings of the Society for Experimental Biology and Medicine 168:290-299 (1981).

3. Clark, D.A., Sweeney, G., Safe, S., Hancock, E., Kilbourn, D.G., and Gauldie, J. Cellular and genetic basis for suppression of cytotoxic T-cell generation by haloaromatic hydrocarbons. Immunopharmacology 6:143-153 (1983).

4. Dean, J.H., Lauer, L.D., House, R.V., Murray, M.J., and Ward, E.C. Effect of 2,3,7,8-tetrachlorodibenzo-p-dioxin exposure on immune parameters in adult mice (in preparation, 1984).

5. Dean, J.H., Luster, M.I., and Boorman, G.A. Immunotoxicology. In: Immunopharmacology, Vol. 4 (P. Sirois and M. Rola-Pleszczynski, Eds.), Elsevier/North-Holland Biomedical Press, Amsterdam, pp. 349-397 (1982).

6. Dean, J.H., Luster, M.I., Boorman, G.A., Chae, K., Lauer, L.D., Luebke, R.W., Lawson, L.D., and Wilson, R.E. Assessment of immunotoxicity induced by the environmental chemicals 2,3,7,8-tetrachlorodibenzo-p-dioxin, diethylstilbesterol and benzo(a)pyrene. In: Advances in Immunopharmacology (J. Hadden, L. Chedid, P. Mullen, and F. Spreafico, Eds.), Pergamon Press, Oxford, pp. 37-50 (1981).

7. Faith, R.E., Luster, M.I., and Vos, J.G. Effect on immunocompetence by chemicals of environmental concern. In: Annual Reviews in Biochemical Toxicology, Vol. 2 (E. Hodgson, J.R. Bend, and R.M. Philpot, Eds.), Elsevier/North-Holland Biomedical Press, Amsterdam, pp. 173-211 (1980).

8. Faith, R.E. and Moore, J.A. Impairment of thymus-dependent immune functions by exposure of the developing immune system to 2,3,7,8-tetrachlorodibenzo-p-dioxin (TCDD). Journal of Toxicology and Environmental Health 3:451-464 (1977).

9. Greenlee, W.F. Molecular mechanisms of immunosuppression induced by 12-o-tetradecanoylphorbol-13-acetate and 2,3,7,8-tetrachlorodibenzo-p-dioxin. In: Toxicology of the Immune System (J. Dean, A. Munson, M. Luster, and H. Amos, Eds.), Raven Press, New York, 1984 (in press).

10. Hay, A. Dioxin as a health hazard. Nature 283:229-230 (1980).

11. Hinsdill, R.D., Couch, D.C., and Speirs, R.S. Immunosuppression in mice induced by dioxin (TCDD) in feed. Journal of Environmental Pathology and Toxicology 3:401-425 (1980).

12. Homberger, E., Reggiani, G., Sambeth, J., and Wipk, H.K. The Seveso Accident: Its nature, extent, and consequence. Annals of Occupational Hygiene 22:327-370 (1979).

13. Kimbrough, R.D. (Ed.) Halogenated Biphenyls, Terphenyls, Naphthalenes, Dibenzodioxins and Related Products, Elsevier/North-Holland Biomedical Press, New York (1980).

14. Lathrop, G.D., Wolfe, W.H., Albanese, R.A., and Moynahan, P.M. An epidemiologic investigation of health effects in Air Force personnel following exposure to herbicides. USAF, Brooks Air Force Base, Texas (1984).

15. Luster, M.I., Boorman, G.A., Dean, J.H., Harris, M.W., Luebke, R.W., Padarathsingh, M.L., and J.A. Moore. Examination of bone marrow, immunologic parameters and host susceptibility following pre- and postnatal exposure to 2,3,7,8-tetrachlorodibenzo-p-dioxin (TCDD). International Journal of Immunopharmacology 2:301-310 (1980).

16. Luster, M.I., Clark. G., Lawson, L.D., and Faith, R.E. Effects of brief in vitro exposure to 2,3,7,8-tetrachlorodibenzo-p-dioxin (TCDD) on mouse lymphocytes. Journal of Environmental Pathology and Toxicology 2:965-977 (1979).

17. Mantovani, A., Vecchi, A., Luini, W., Sironi, M., Candiani, G.P., Spreafico, F., and Garattini, S. Effect of 2,3,7,8-tetrachlorodibenzo-p-dioxin on macrophage and natural killer cell-mediated cytotoxicity in mice. Biomedicine 32:200-204 (1980).

18. McConnell, E.E. Acute and chronic toxicity, carcinogenesis, reproduction, teratogenesis, and mutagenesis in animals. In: Halogenated Biphenyls, Terphenyls, Naphthalenes, Dibenzodioxins and Related Products (R.D. Kimbrough, Ed.), Elsevier/North-Holland Biomedical Press, New York, pp. 109-150 (1980).

19. Neal, R.A. Biological effects of 2,3,7,8-tetrachlorodibenzo-p-dioxin in experimental animals. Paper presented at this Symposium.

20. Nicholson, W.J. and Moore, J.A. (Eds.). Health Effects of Halogenated Aromatic Hydrocarbons. Annals of the New York Academy of Sciences 320:1-730 (1979).

21. Price, J.K. Toxic effects of in vivo dioxin-chlorophenol exposure on human leucocyte functions. Unpublished (1984).

22. Poland, A. and Glover, E. 2,3,7,8-Tetrachlorodibenzo-p-dioxin: Segregation of toxicity with the Ah locus. Molecular Pharmacology 17:86-94 (1980).

23. Poland, A., Greenlee, W.F., and Kende, A.S. Studies on the mechanism of action of the chlorinated dibenzo-p-dioxins and related compounds. Annals of the New York Academy of Sciences 320:214-230 (1979).

24. Poland, A. and Knutson, J.C. 2,3,7,8-Tetrachlorodibenzo-p-dioxin and related halogenated aromatic hydrocarbons: Examination of the mechanisms of toxicity. Annual Review Pharmacology and Toxicology 22:517-554 (1982).

25. Reggiani, G. Acute human exposure to TCDD in Seveso, Italy. Journal of Toxicology and Environmental Health 6:27-43 (1980).

26. Sharma, R.P. and Gehring, P.J. Effects of 2,3,7,8-tetrachlorodibenzo-p-dioxin (TCDD) on splenic lymphocyte transformation in mice after single and repeated exposures. Annals of the New York Academy of Sciences 320:487-497 (1979).

27. Thigpen, J.E., Faith, R.E., McConnell, E.E., and Moore, J.A. Increased susceptibility to bacterial infection as a sequela to exposure to 2,3,7,8-tetrachlorodibenzo-p-dioxin. Infection and Immunity 12:1319-1324 (1975).

28. Thomas, P.I. and Hinsdill, R.D. The effect of perinatal exposure to tetrachlorodibenzo-p-dioxin on the immune response of young mice. Drug and Chemical Toxicology 2:77-98 (1979).

29. Van Logten, M.J., Gupta, B.N., McConnell, E.E., and Moore, J.A. Role of the endocrine system in the action of 2,3,7,8-tetrachlorodibenzo-p-dioxin (TCDD) on the thymus. Toxicology 15:135-144 (1980).

30. Vecchi, A., Mantovani, A., Sironi, M., Luini, W., Spreafico, F., and Garattini, S. The effect of acute administration of 2,3,7,8-tetrachlorodibenzo-p-dioxin (TCDD) on humoral antibody production and cell-mediated activities in mice. Archives of Toxicology 4:163-165 (1980).

31. Vecchi, A., Sironi, M., Canegrati, M.A., Recchia, M., and Garattini, S. Immunosuppressive effects of 2,3,7,8-tetrachlorodibenzo-p-dioxin in strains of mice with different susceptibility to induction of aryl hydrocarbon hydroxylase. Toxicology and Applied Pharmacology 68:434-441 (1983).

32. Vos, J.G. Immune suppression as related to toxicology. CRC Critical Reviews in Toxicology 5:67-101 (1977).

33. Vos, J.G., Faith, R.E., and Luster, M.I. Immune alterations. In: Halogenated Biphenyls, Terphenyls, Naphthalenes, Dibenzodioxins and Related Products (R.D. Kimbrough, Ed.), Elsevier/North-Holland Biomedical Press, New York, pp. 241-266 (1980).

34. Vos, J.G., Kreeftenberg, J.G., Engel, H.W.B., Minderhoud, A., and Van Noorle Jansen, L.M. Studies on 2,3,7,8-tetrachlorodibenzo-p-dioxin-induced immune suppression and decreased resistance to infection: Endotoxin hypersensitivity, serum zinc concentrations and effect of thymosin treatment. Toxicology 9:75-86 (1978).

35. Vos, J.G. and Moore, J.A. Suppression of cellular immunity in rats and mice by maternal treatment with 2,3,7,8-tetrachlorodibenzo-p-dioxin. International Archives of Allergy and Applied Immunology 47:777-794 (1974).

36. Vos, J.G., Moore, J.A., and Zinkl, J.G. Effect of 2,3,7,8-tetrachlorodibenzo-p-dioxin on the immune system of laboratory animals. Environmental Health Perspectives 5:149 (1973).

ALTERATION OF RAT HEPATIC PLASMA MEMBRANE FUNCTIONS BY TCDD (2,3,7,8-TETRACHLORODIBENZO-*p*-DIOXIN)

Fumio Matsumura, Ph.D.
David Brewster, M.S.
Burra V. Madhukar, Ph.D.
David Bombick, M.S.

Pesticide Research Center
Michigan State University
East Lansing, Michigan 48824

This paper closely follows Dr. Matsumura's presentation during the symposium. Adapted, with permission, from Archives of Environmental Contamination and Toxicology 13, 509-515 (1984).

From Public Health Risks of the Dioxins, proceedings of a symposium held on October 19-20, 1983 at The Rockefeller University, New York, City. Edited by William W. Lowrance. Copyright, The Rockefeller University, 1984. Published by William Kaufmann, Los Altos, California.

2,3,7,8-Tetrachlorodibenzo-*p*-dioxin (TCDD) is a serious environmental pollutant (e.g., Blair, 1973). It was first discovered as a toxic contaminant in the herbicide 2,4,5-T. This included Agent Orange used in Vietnam. Subsequently low level residues of TCDD have been detected in soil, fish, and industrial and municipal fly ashes in various locations throughout the United States (e.g., Bumb *et al.*, 1980). It is said to be the most toxic small size chemical known to man, its LD_{50} to guinea pigs being in the range of 0.6 ug/kg (single oral dose) (Schwetz *et al.*, 1973). It is also teratogenic, carcinogenic and acnegenic (Kociba *et al.*, 1978). However, its toxic manifestations vary from species to species.

Because of such diverse effects of TCDD the search for the biochemical mechanisms of its toxic action has been difficult. The most significant advance made in this regard has been the characterization of hepatic induction in rodent species (Poland and Kende, 1976; Nebert, 1979). It was originally proposed by Poland (Poland and Glover, 1974) that TCDD first binds with a specific cytosolic receptor, is transferred into the nucleus, and mediates the expression of a number of microsomal and cytosolic enzymes. Yet, it is not a prominent biochemical response in some susceptible species such as the guinea pig (Hook *et al.*, 1975; Gupta *et al.*, 1973).

Some of the most consistent symptoms caused by TCDD among all experimental animals are involution of the thymus (Gupta *et al.*, 1973; Harris *et al.*, 1973; Vos *et al.*, 1973), loss of body weight, and the subsequent appearance of "wasting" syndrome at toxic doses (e.g., Seefeld, 1982). Intensive research in this regard turned up no clue for such symptoms (McConnell and Moore, 1978; Neal *et al.*, 1979). Other investigations to find interspecies differences among guinea pigs, rats and hamsters, etc. in the metabolism, disposition and cytosolic receptor binding of TCDD have turned out modest differences, but they may not be significant enough to explain the enormous species difference in susceptibility to this toxicant (Gasiewicz *et al.*, 1983).

Recently, using electrophoresis we found that protein profiles of hepatic plasma membrane from TCDD-treated rats were different from those of untreated controls (Brewster *et al.*, 1982). In this report we present evidence that *in vivo* exposure to TCDD affects a number of physiologically important components of the rat liver plasma membrane.

MATERIALS AND METHODS

Throughout the study young male Sprague-Dawley rats were used and fed Purina Laboratory Chow, *ad libitum*. TCDD was dissolved in corn oil with acetone (9:1 ratio) and used for intraperitoneal (i.p.) injection. Control rats received the same volume of corn oil-acetone. After specified time periods they were sacrificed, and the hepatocyte plasma membrane was isolated according to the method of Yunghans and Morre (1973).

ATPases were assayed by the method described by Matsumura and Clark (1980). Modifications were: 25 ug plasma membrane protein in 0.1 ml sucrose (0.25 M) added to 0.9 ml reaction buffer containing 10 ul 2,4-dinitrophenol (10 mM) was preincubated at 37°C (10 minutes). (Gamma-^{32}P) ATP was added (final concentration 10 mM) and the mixture incubated at 37°C for 10 min. The reaction was stopped with the addition of 300 ul ice cold trichloroacetic acid (10%) followed by 100 ul H_2O containing 1 mg bovine serum albumin and 136 mg KH_2PO_4. The solution was centrifuged (5 min, 1000 x g) and the clear supernatant decanted and mixed with 200 ug activated charcoal. The sides of the tube were rinsed with 200 ul ethanol, and after a second 5 minute spin 0.5 ml of the supernatant was mixed with aqueous scintillation cocktail for liquid scintillation counting. Buffers for determining various ATPase activities contained 30 mM imidazole (pH 7.1) with the following ion combinations: Na-K ATPase, 120 mM NaCl, 20 mM KCl, and 5 mM $MgCl_2$; Mg-ATPase, 120 mM NaCl, 5 mM $MgCl_2$; Basal medium for Ca-ATPase, 120 mM NaCl, 20 mM KCl, 2 mM $CaCl_2$, 0.5mM EGTA; and for Ca-ATPase, 120 mM NaCl, 20 mM KCl and 0.5 mM EGTA and 2 mM $CaCl_2$. Protein kinase activity was determined by the method of Corbin and Reiman (1974) with the following modifications: plasma membranes (50 ug protein) in 0.25 M sucrose were added to 50 ul reaction buffer containing 1 ml 50 mM potassium phosphate (pH 6.8), 1 ml 30 mg/ml histone (Sigma Chem. Co., II A-S), and 1 ml 1 mM (gamma-^{32}P)-ATP in 18 mM magnesium acetate. For determining c-AMP dependent protein kinase activity 6 nmoles c-AMP in 1 ml of water was added to the reaction buffer. After incubating for 10 min at 30°C the reaction was stopped with 3 ml cold trichloroacetic acid (10%) and 100 ul water, containing 100 mg bovine serum albumin and 136 mg KH_2PO_4, was added. The tubes were allowed to stand for 5 min for complete protein precipitation, spun for 5 min (1000 x g), the supernatant decanted, and the pellet redissolved with 0.5 ml NaOH (0.2N). Reprecipitation with trichloroacetic acid and a second centrifugation followed. After repeating this washing procedure once more (total of 2 times) the final pellet was resuspended with 0.3 ml formic acid, of which 0.1 ml was used for liquid scintillation counting.

Gamma-glutamyl transpeptidase activity was assessed by the method of Tate and Meister (1974). Additional information on this technique is: 100 ug membrane protein (suspended in 0.25 M sucrose) in 0.1 ml tris-HCl (0.01M)-NaCl (0.15M), pH 8.0 was mixed with the assay solution and incubated 15 min at 37°C. After stopping the reaction with 0.1 ml glacial acetic acid, spinning for 5 min (1000 x g) the absorbance of p-nitroanaline was determined at 410 nm with a Varian Double Beam Spectrophotometer.

Binding of [^{125}-I]-insulin and [^{125}-I]-epidermal growth factor (EGF) was studied essentially following the method of O'Keefe _et al._ (1974). Other experimental details are: plasma membranes (50 ug protein) were suspended in 0.2 ml of Kreb's Ringer bicarbonate buffer (pH 7.4) containing 0.1% bovine serum albumin (BSA) and incubated with or without the native ligands (insulin 2.0 ug, EGF 0.5 ug) for 10 min at 24°C before the addition of 0.25 ng of the labeled ligand (insulin Sp.Act. 100 uCi/ug, EGF, Sp. Act. 150-200 uCi/ug). All the tubes were incubated for an additional period of 20 min. At the end of the incubation the reaction mixture was diluted with 3.0 ml of chilled Kreb's Ringer bicarbonate buffer containing 0.5% BSA and quickly filtered through a Millipore filter (type HAWP, 0.45u) under vacuum. The filters were washed twice with 5.0 ml aliquots of the same buffer, air dried and counted for radioactivity.

Binding of ^{125}I-glucagon was determined using a similar procedure for insulin with the following modifications: the reaction medium contained 20 mm tris-HCl with 1% BSA, and 0.2 ng of ^{125}I-glucagon (New England Nuclear Sp. Act. 218 uCi/ug). Parallel tubes in duplicate were pre-incubated for 10 min with 0.2 ug of native glucagon before the addition of the labeled ligand to determine non-specific binding. The wash medium contained 20 mM tris with 1% BSA. The filters used were cellulose acetate (Millipore EGWP 0.2u).

^{3}H-PGE$_1$ (prostaglandin E$_1$) binding to liver plasma membrane was determined by adopting the filter method described above for insulin, EGF and glucagon with modifications as follows: the reaction medium of 0.2 ml contained 150 ug of plasma membrane protein in 150 mM NaCl, 10 mM PIPES [Piperazine N,N'-bis(2-ethane sulfonic acid)] buffer, pH 6.5. The membranes were pre-incubated for 10 min at 37°C with or without 1 nmole unlabeled PGE$_1$. ^{3}H-PGE$_1$ (Amersham Corporation, Sp. Act. 40 Ci/mmol) 1 pmole was added to the reaction mixture and the tubes were further incubated for 20 min at the same temperature. At the end of incubation, 3.0 ml of cold NaCl-PIPES buffer with 0.1% BSA was added to each tube and quickly filtered over Millipore filter (HAWP, 0.45u). The filters were washed twice with 5 ml aliquots of the same buffer and the radioactivity counted as described above. In all cases, specific binding was calculated by subtracting the amount of radioactivity bound in the presence of native ligand from that in the absence of it.

Binding of [^{3}H]-concanavalin A (Con A) was studied by the method originally used by Chandramouli *et al.* (1977). The following modifications were made: 25 ug of plasma membrane protein were suspended in a 0.2 ml volume of 0.1 M tris-CHl buffer, pH 7.4 containing 0.1% BSA. Duplicate tubes were preincubated with or without *a*-methyl D-mannopyranoside (*a*-mm) for 10 min at 37°C before the addition of 0.25 ug of ^{3}H-Con A (Specific Activity 25-45 Ci/m mol, New England Nuclear, Boston, Mass.). At the end of an additional 10 min of incubation at the same temperature, 8.0 ml of cold tris-HCl (0.1 M, pH 7.4) containing 0.5% BSA was added to each tube and the contents were quickly filtered over a Millipore filter (HAWP, 0.45u). The filter was washed with an additional 8.0 ml of the same buffer and the radioactivity was counted as described above. Specific binding was defined as the difference between the amount of radioactivity in the presence and in the absence of *a*-mm.

RESULTS

To study the biochemical characteristics of the plasma membrane the rats were first treated with 25 ug/kg of TCDD (single dose) through intraperitoneal injection (i.p.), and their hepatic plasma membrane was isolated after 10 days as before (Peterson *et al.*, 1979). The difference in the yields of plasma membrane between TCDD-treated and control rats was not significant. One gram wet weight liver yielded on average 240 ± 102 (7 determinations, mean ± S.E.) and 226 ± 75 ug (9 determinations) of proteins from control and TCDD-treated rats, respectively.

However, when these plasma membrane preparations were analyzed through SDS-polyacrylamide gel electrophoresis, qualitative differences in protein composition became apparent. In the electropherotogram shown in Figure 1, one can recognize that the most significant effect of *in vivo* administered TCDD was observed in the preparations obtained from rats sacrificed at day 10 and 20 after TCDD treatment.

Figure 1. SDS-polyacrylamide slab gel electrophoresis patterns of hepatic plasma membrane from the rats. (A) treated with 25 ug/kg (single i.p.) of TCDD 40 days after treatment. (B) Same after 20 days. (C) Same after 10 days. (D) Same after 2 days. (E) Untreated control, day 2. The mortality at day 20 was 0 and at day 40 was 20 to 30%. Standard reference proteins used were myosin (120), bovine albumin (66), egg albumin (45), trypsinogen (24 k), gamma-lactoglobulin (18.4) and lysozyme (14). Numbers in parentheses are the subunit size in kilodaltons.

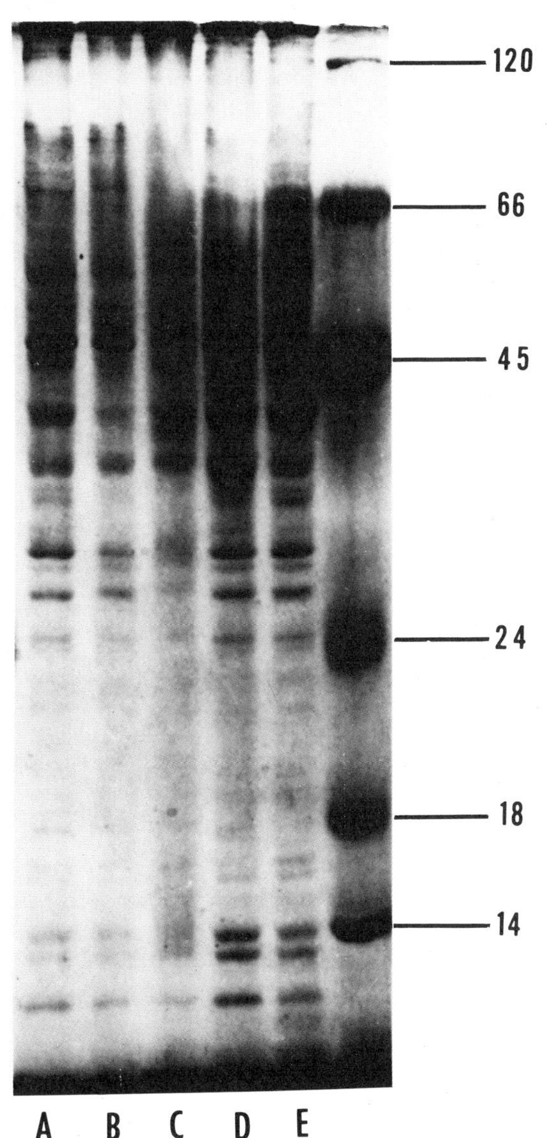

Some of the bands between 14 and 30 K dalton are completely abolished by TCDD-treatment. The densitometric measurement of these bands (data not shown) confirms the visual conclusion. In this case the band at 48 K which is a structural protein comprising several percents of the membrane protein served as a good internal standard (see Brewster et al., 1982) as this band intensity remained constant throughout the treatment period.

To study the qualitative nature of the altered plasma membrane the activities of several membrane bound enzymes and receptor proteins were examined.

Table 1. Difference[a] in various enzyme activities between the hepatic plasma membrane preparation from untreated control and _in vivo_ TCDD-treated rats. The results are expressed as mean ± standard error and the number in parentheses indicates the number of different membrane preparations tested. Each preparation was tested twice.

Enzyme	Enzyme Activities (nmoles product/mg protein/hr) Control	TCDD-treated[b]
Na-K ATPase[c]	1496 ± 142 (3)	890 ± 178 (3)***
Mg ATPase[c]	1820 ± 10 (2)	1110 ± 149 (2)**
Ca ATPase[c]	608 ± 65 (4)	340 ± 140 (8)***
Gamma-glutamyl transpeptidase[d]	887 ± 231 (4)	658 ± 35 (3)*
Protein kinase[e]		
in the absence of c-AMP	55 ± 31 (6)	147 ± 67 (7)**
in the presence of c-AMP	58 ± 25 (4)	217 ± 110 (7)**

[a]Data are analyzed using the student "t" test: * $P<0.1$, ** $P<0.05$, *** $P<0.01$.

[b]TCDD at 25 ug/kg single dose intraperitoneal injection. Plasma membrane collected at 10 days post-treatment.

[c]n moles P_i liberated/mg plasma membrane protein/hr at pH 7.1, 37°C.

[d]n moles p-nitroaniline liberated/mg plasma membrane protein/hr at pH 8.0, 37°C.

[e]n moles P_i incorporated/mg plasma membrane protein/hr at pH 8.0, 37°C.

The differences in Na-K and Ca ATPase are highly significant. The difference in gamma-glutamyl transpeptidase activity between TCDD-treated and untreated rats was only marginally significant. The level of protein kinase in the plasma membrane from treated rats was significantly higher than that of the control, indicating that TCDD treatment does not always cause a reduction in enzyme activities. In this case both c-AMP stimulated and nonstimulated protein kinases from the TCDD-treated animals showed higher enzyme activities than those from the controls.

Table 2. Effect of *in vivo* TCDD treatment on ligand binding to cell-surface membrane receptors in the rat liver. TCDD was intraperitoneally administered to rats at indicated doses. Plasma membrane was fractionated 10 days after treatment. Ligand binding was studied as described under Materials and Methods. The results are expressed as mean ± S.E. and the number of animals tested is given in parentheses.[a]

Ligand	Dose (ug/kg single i.p.)	Specific Binding	
		Control	TCDD-treated
^{3}H-Con A[b]	25	15.2 ± 2.5(4)	11.7 ± 0.4(3)**
^{125}I-EGF[c]	25	32.0 ± 8.6(5)	3.8 ± 0.3(3)***
^{125}I-Glucagon[c]	25	43.5 ± 10.1(6)	43.7 ± 6.8(4)***
	115		4.4 ± 4.1(8)
^{125}I-Insulin[c]	25	8.1 ± 1.2(7)	11.7 ± 0.9(3)***
^{3}H-Prostaglandin E_1[d]	25	3.6 ± 0.3(3)	3.3 ± 2.3(6)

[a]Data are analyzed using the student "t" test; ** $P<0.05$, ***$P<0.01$.

[b]ng of ^{3}H-Con A bound/25 ug protein/10 min.

[c]pg of ^{125}I-EGF, glucagon- or insulin-bound/50 ug protein/20 min.

[d]pg of ^{3}H-PGE$_1$ bound/150 ug protein/20 min.

As for receptor activities (Table 2) EGF-binding was most severely affected at 25 ug/kg (single i.p.) dose of TCDD treatments. In agreement with our previous observation (Brewster *et al.*, 1982) Con A binding was also found to be reduced. On the other hand, the effects of TCDD on glucagon receptors was significant only at a high dose (115 ug/kg). The effect on insulin binding was unusual in that at low doses it was stimulated and at the high dose it was inhibited (data not presented). Under our experimental conditions TCDD treatment caused no appreciable change in prostaglandin E_1 binding.

The time course of TCDD effects was studied following a single i.p. dose (25 ug/kg, Figure 2).

Figure 2. Time course of changes in specific binding of insulin or Con A to liver plasma membranes and body weight changes of TCDD-treated rats relative to untreated control rats. Rats were given a single i.p. dose of TCDD at 25 ug/kg and plasma membranes prepared at 2, 10, 20 and 40 days after treatment. Control rats received an appropriate volume of the vehicle (corn oil:acetone, 9:1). Binding was expressed as pg of ligand bound/50 ug protein for ^{125}I-insulin, or ng of ^{3}H-Con A bound/25 ug protein. Changes in the body weight of the TCDD-treated rats were expressed as % of body weight of controls at each time point. Values for controls are: 108 ± 2, 140 ± 8, 170 ± 15 and 206 ± 18 g. at 2, 10, 20 and 40 days respectively. Each point is a mean ± S.E. of 4 to 8 animals. The data are analyzed using student "t" test. Asterisk designations are same as in Table 1.

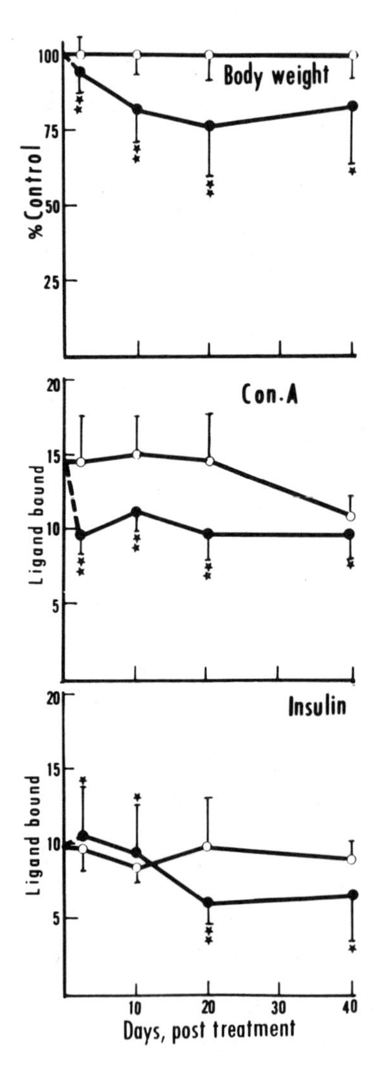

During the 40 day observation period TCDD-treated rats gained consistently less body weight than did the control rats. Insulin binding was unaffected at the beginning but by day 20 after treatment it was significantly reduced (P<0.1). Con A binding was continuously suppressed.

The time course of TCDD effects on the membrane enzyme activities were studied at the same time and dose regimen.

Figure 3. Time course of changes in plasma membrane associated enzyme activities in the rat liver as a result of in vivo TCDD exposure. TCDD was administered as a single i.p. dose at 25 ug/kg. Hepatic plasma membrane was isolated at indicated days of post treatment and the various enzyme activities were assayed in the isolated membrane as described under Materials and Methods. Values are expressed as % control activities and are represented (where indicated) by mean ± S.E. of at least 3 different membrane preparations. Control values for the enzymes assayed at 2, 10, 20 and 40 days respectively are: Total ATPase, 972, 1479 ± 142, 1768 and 1137 nmoles of Pi liberated/mg protein/hr; Mg-ATPase: 1211, 1820 ± 10, 1076, and 1230 nmoles of Pi/mg protein/hr; Ca-ATPase, 564 ± 119, 609 ± 65, 603 ± 105, and 638 ± 152 nmoles Pi/mg protein/hr; and gamma-glutamyl transpeptidase (gamma-GT) 884 ± 116, 887 ± 231, 935 ± 223 and 666 ± 68 nmoles of p-nitroaniline produced mg protein/hr at 37°C and pH 8.0. It should be noted that an apparent recovery is seen on day 40 which reflects only the levels of these biochemical parameters in the surviving population. By day 40, 20 to 30% of the population died at this dose (25 ug/kg).

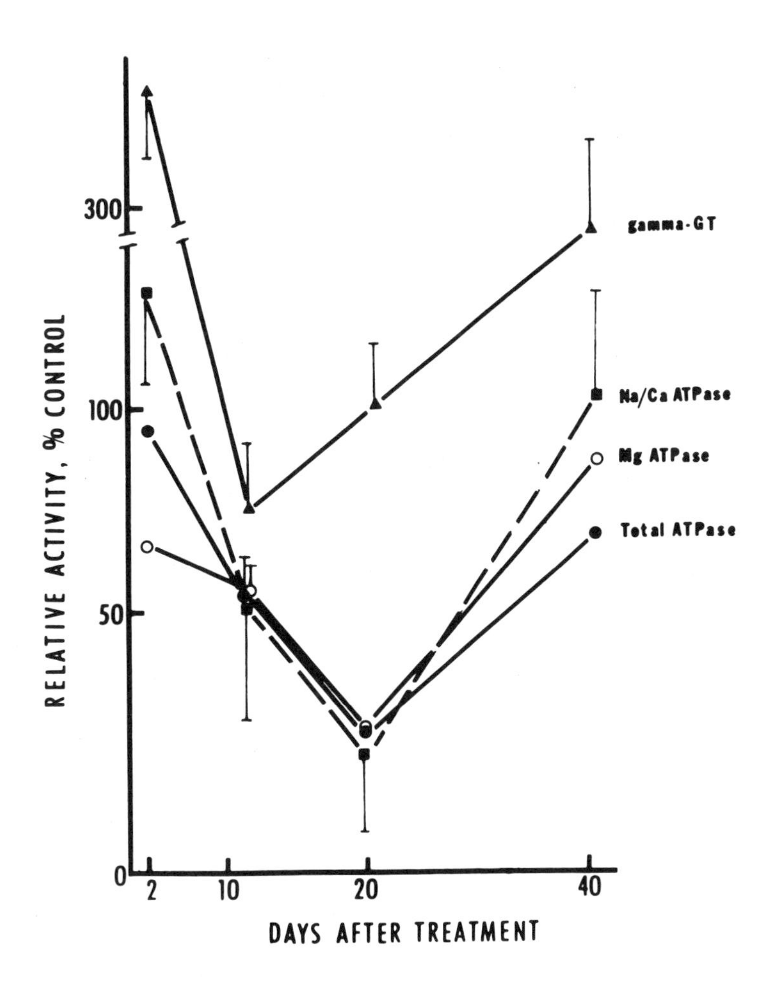

The results shown in Figure 3 clearly indicate that, with the exception of gamma-GT activities, these enzyme activities follow a reduction pattern similar to that of the membrane bound receptors: The decline becomes significant at day 10 and reaches the minimum level at day 20 followed by an apparent recovery. There is a likely possibility that this apparent recovery does not represent a true biochemical recovery, since by day 40 approximately 20 to 30% of the treated rats had died, while at other times no mortality was observed.

In terms of the percentage of reduction, it may be concluded that among the criteria examined EGF binding was the most sensitive parameter to *in vivo* TCDD treatment (Table 2). It is also significant to note that this phenomenon is already observed on the second day after treatment (Figure 2), as in the case of the body weight loss.

As mentioned above, in the case of insulin binding (Figure 2), TCDD's effect was not significant on day 10 of treatment. However, it was noted that at 25 ug/kg single i.p. some rats showed loss in body weight, while others did not. Therefore, an attempt was made to correlate body weight changes with the level of insulin binding to plasma membrane from individual rats at 20 days after treatment. In the data summarized in Figure 4 all individual values from treated (25 and 115 ug/kg) and control rats were pooled. Analysis of coefficient of correlation indicates that there is a significant relationship between the reduction in insulin binding and changes in weight.

Figure 4. Correlation between ^{125}I-insulin specific binding to liver plasma membrane and body weight changes in rats treated with a single i.p. dose of TCDD at 25 ug/kg or 115 ug/kg at 20 days post-treatment. Specific binding of insulin for plasma membrane preparations from individual rats was determined as described under Materials and Methods. Body weight levels of individual rats at 20 days after treatment are plotted against specific insulin binding. The curve was generated by computer using the points of all preparations of TCDD treated rats (◘ for 25 ug/kg, ■ 115 ug/kg) and control rats which were treated only with the vehicle (◉, corn oil:acetone, 9:1). Data are analyzed using least squares regression, coefficient of correlation R=0.8288.

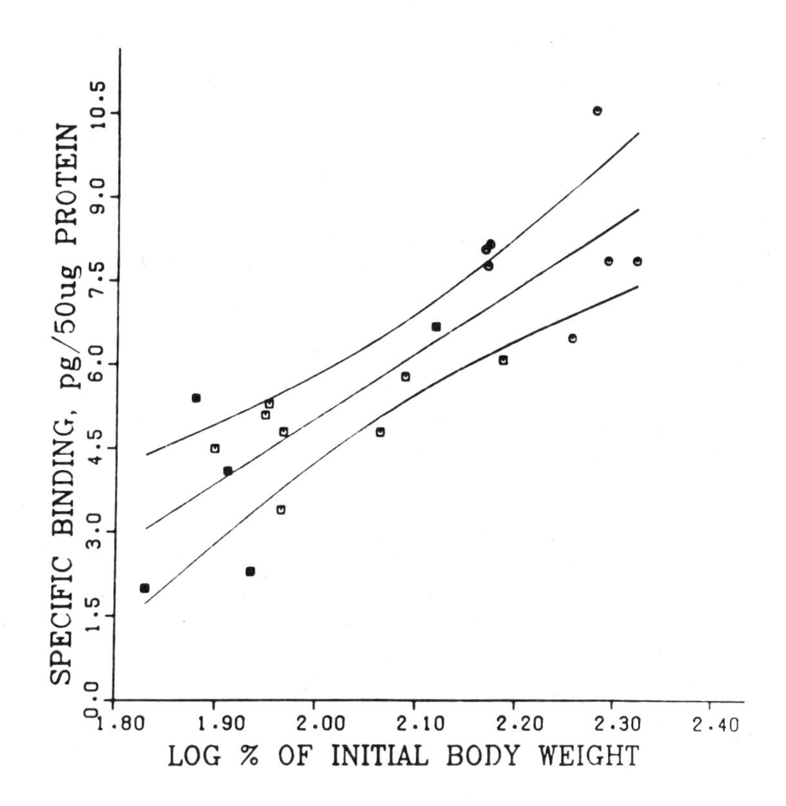

At this stage a question may be raised as to whether this phenomenon of induced changes in plasma membrane functions is associated with any toxicants which cause general stress or is confined to TCDD. To answer this question the rats were treated (all i.p. daily for 10 days with 6:1 corn oil and acetone) with 3-methylcholanthrene (20 mg/kg), Aroclor 1242 (50 mg/kg), phenobarbital (120 mg/kg), DDT (0.3 mg/kg) and Firemaster B-6 (i.e., PBB, 50 mg/kg). At the end of the *in vivo* treatment the animals were sacrificed, the hepatic plasma membrane was isolated, and the extent of Con A binding was studied. The levels of specific binding as expressed in terms of ng/50 ug plasma membrane protein were: control 17.0 ± 2.0, 3-methylcholanthrene 21.0 ± 4.9, Aroclor 1242 18.4 ± 2.1, phenobarbital 10.6 ± 1.8, DDT 13.9 ± 1.2 and Firemaster B-6 19.5 ± 6.6 (mean ± standard deviation, 3 animals except for DDT and Firemaster experiments where 2 animals were used).

Only the results of the phenobarbital experiment were statistically different from the control value (at $P \geq 0.05$). These results, therefore, indicate that such changes in the plasma membrane are caused only by rather specific chemicals.

DISCUSSION

In the current investigation we have found that the activities of several enzymes and receptors in the plasma membrane in the TCDD-treated rat liver were significantly different from those in the untreated control rat liver.

It must be pointed out that the purpose of this investigation was to survey how widely such TCDD-induced membrane changes are occurring. This follows our earlier observation (Brewster *et al.*, 1982) that the levels of a number of proteins of the rat liver plasma membrane were altered as a result of TCDD poisoning. To this end we have surveyed 11 parameters and found that nine of them have been affected by TCDD treatment. An inevitable conclusion is that extensive biochemical changes are taking place in the rat hepatic plasma membrane as a result of TCDD exposure. Consequently one could surmise that there may be many other changes occurring that have not been recognized in this study. Also, if any changes occur in critical enzymatic functions or receptor sites, serious toxicological consequences may be expected.

On the other hand, as to the apparent correlation between the *in vitro* changes and the *in vivo* observations (e.g., the reduction in insulin binding and the body weight loss, or the similarities in the time course of the changes of these and other parameters), one must not draw the simple conclusion that these phenomena are directly related. However, there is no question about the importance of the roles of insulin receptors in the process of nutritional homeostasis which is intimately involved in body weight maintenance. Serious consideration must be given, therefore, to the possibility that TCDD-triggered changes in insulin receptor activity cause severe nutritional imbalance which ultimately results in body weight loss.

As to the cause of alteration of the plasma membrane proteins we have no explanation to offer at present. Since TCDD is known to bind with the cytosolic receptor (Poland *et al.*, 1976; Poland and Kende, 1976; Nebert, 1979) and is transported into the nucleus, it may be reasonable to assume that the changes in the plasma membrane enzyme and receptor activities observed in this work are triggered by induced DNA pleiotropic responses rather than by TCDD's direct interaction with the plasma membrane. There is evidence that the reduction of Na-K ATPase activity, as a result of *in vivo* administration of TCDD, is not due to a direct interaction of this chemical with the enzyme itself (Peterson *et al.*, 1979), since the amount of TCDD found in the plasma membrane after *in vivo* administration is not enough to cause inhibition of this enzyme *in vitro*.

While this investigation does not indicate that the cause of these changes is causally related to TCDD's toxic action, the phenomenon of changes in plasma membrane functions as a result of TCDD poisoning merits further consideration. For instance, some of the altered enzymes and receptors found in the current investigation are known to carry out very important physiological and biochemical functions, such as Ca-ATPase and Na-K ATPase that are involved in the transport of Ca^{2+} and Na^{+} across the plasma membrane. Certainly, insulin receptor is a vital system that controls carbohydrate and lipid metabolism and homeostasis. Furthermore, many physiological homeostatic mechanisms depend on cellular surface functions; therefore, alterations in cell surface could conceivably lead to some toxic manifestations of TCDD. Since cell surface membranes vary according to tissue, species and sex, it is reasonable to expect different toxic responses to TCDD in different species.

ACKNOWLEDGMENTS

Supported by Michigan Agricultural Experiment Station (Journal Article No. 11219), Michigan State University, and by research grant ES01963 from the National Institute of Environmental Health Sciences, Research Triangle Park, North Carolina.

ABSTRACT

In vivo administration of 2,3,7,8-tetrachlorodibenzo-p-dioxin to rats was found to produce significant alterations in Na-K, Ca, Mg ATPase and protein kinase activity, and receptor binding activity to insulin, concanavalin A, glucagon and epidermal growth factor in the canaliculi-rich plasma membrane fraction of the hepatocytes. Some of the changes are dose-related, and generally follow the time course of TCDD poisoning in the rat (e.g., the loss of body weight). Such TCDD-induced changes in membrane proteins and subsequent alteration in membrane functions appear to be extensive.

REFERENCES

Blair, E.H. (ed.). 1973. Chlorodioxins-Origin and Fate. (Advances in Chemical Series 120) American Chemical Society, Washington, D.C.

Brewster, D.W., Madhukar, B.V. and Matsumura, F. 1982. Influence of 2,3,7,8-TCDD on the protein composition of the plasma membrane of hepatic cells from the rat. Biochemical and Biophysical Research Communications 107:68-74.

Bumb, R.R., Crummett, W.B., Cutie, S.S., Gledhill, J.R., Hummel, R.H., Kagel, R.O., Lamparski, L.L., Luoma, E.V., Miller, D.L., Nestrick, T.J., Shadoff, L.A., Stehl, R.H. and Woods, J.S. 1980. Trace characteristics of fire. A source of chlorinated dioxins. Science 210:385-394.

Chandramouli, V., Williams, S., Marshall, J.S. and Carter, Jr., J.R. 1977. Cell surface changes in diabetic rats. Studies of lectin binding to liver cell plasma membranes. Biochimica et Biophysica Acta 465:19-33.

Corbin, J.D. and Reiman, E.M. 1974. Assay of c-AMP dependent protein kinases. In: Hardman, J.G., and O'Malley, B.W. (eds.). Methods in Enzymology. Academic Press, New York, 38:287-290.

Gasiewicz, T.A., Olson, J.R., Gieger, L.E. and Neal, R.A. 1983. Absorption, distribution and metabolism of 2,3,7,8-tetrachlorodibenzo-p-dioxin (TCDD) in experimental animals. In: Tucker, R.E., Young, A.L., and Gray, A.P. (eds.). Human and Environmental Risks of Chlorinated Dioxins and Related Compounds. Plenum Press, New York, 495-525.

Gupta, B.N., Vos, J.G., Moore, J.A., Zinkl, J.G. and Bullock, B.C. 1973. Pathological effects of 2,3,7,8-tetrachlorodibenzo-p-dioxin in laboratory animals. Environmental Health Perspectives 5:125-140.

Harris, M.W., Moore, J.A., Vos, J.G. and Gupta, A.B.N. 1973. General biological effects of TCDD in laboratory animals. Environmental Health Perspectives 5:101-109.

Hook, G.E.R., Haseman, J.K. and Lucier, G.W. 1975. Induction and suppression of hepatic and extra-hepatic microsomal foreign-compound-metabolizing enzyme systems by 2,3,7,8-TCDD. Chemico-Biological Interactions 10:199-214.

Kociba, R.J., Keyes, D.G., Beyer, J.E., Carreon, R.M., Wade, C.E., Dittenber, D., Kalnins, R., Frauson, L., Park, C.N., Bernard, S., Hummel, R. and Humiston, C.G. 1978. 2,3,7,8-Tetrachloro-dibenzo-p-dioxin (TCDD): Results of a 13-week oral toxicity study in rats. Toxicology and Applied Pharmacology 46:279-303.

Matsumura, F. and Clark, J.M. 1980. ATPases in the axon-rich membrane preparation from the retinal nerve of the squid, Loligo paelei. Comparative Biochemistry and Physiology 66B:23-32.

McConnell, E.E. and Moore, J.A. 1978. The toxicopathology of TCDD. In: Cattabeni, F., Cavallaro, A., and Galli, G. (eds.). Dioxin, Toxicological and Chemical Aspects (Monograph of the Giovanni Lorenzini Foundation), Spectrum Publications, New York, 137-141.

Neal, R.A., Beatty, P.W. and Gasiewicz, T.A. 1979. Studies of the mechanisms of toxicity of 2,3,7,8-tetrachlorodibenzo-p-dioxin (TCDD). Annals of the New York Academy of Sciences 320:204-213.

Nebert, D.W. 1979. Genetic differences in the induction of monooxygenase activities by polycyclic aromatic compounds. Pharmacology and Therapeutics 6:395-417.

O'Keefe, E., Hollenberg, M.D. and Cuatracasas, P. 1974. Epidermal growth factor characteristics of specific binding in membranes from liver, placenta, and other target tissues. Archives of Biochemistry and Biophysics 164:518-526.

Peterson, R.E., Madhukar, B.V., Yang, K.H. and Matsumura, F. 1979. Depression of adenosine triphosphatase activities in isolated liver surface membranes of 2,3,7,8-tetrachlorodibenzo-p-dioxin treated rats: correlation with effects on ouabain biliary excretion and bile flow. Journal of Pharmacology and Experimental Therapeutics 210:275-282.

Poland, A. and Glover, E. 1974. Comparison of 2,3,7,8-tetrachloro-dibenzo-p-dioxin, a potent inducer of aryl hydrocarbon hydroxylase, with 3-methylcholanthrene. Molecular Pharmacology 10:349-359.

Poland, A., Glover, E. and Kende, K.S. 1976. Stereospecific, high affinity binding of 2,3,7,8-tetrachlorodibenzo-p-dioxin by hepatic cytosol. Journal of Biological Chemistry 251:4936-4946.

Poland, A. and Kende, A. 1976. 2,3,7,8-tetrachlorodibenzo-p-dioxin: Environmental contaminant and molecular probe. Federation Proceedings 35:2404-2411.

Schwetz, B.A., Norris, J.M., Sparchu, G.L., Row, V.K., Gehring, P.J. and Emerson, J.L. 1973. Toxicology of chlorinated dibenzo-p-dioxins. Environmental Health Perspectives 5:87-99.

Seefeld, M.D., Corbett, S.W., Keesey, R.E. and Peterson R.E. 1982. Comparison of body weight loss and oxygen consumption in rats treated with 2,3,7,8-tetrachlorodibenzo-p-dioxin (TCDD) and pair-fed control rats. The Toxicologist 2:131 (Abstract 462).

Tate, S.S. and Meister, A. 1974. Interaction of gamma-glutamyl transpeptidase with amino acids, dipeptides, and derivatives and analogs of glutathione. Journal of Biological Chemistry 249:7593-7602.

Vos, J.G., Moore, J.A. and Zinkl, J.G. 1973. Effect of 2,3,7,8-tetrachlorodibenzo-p-dioxin on the immune system of laboratory animals. Environmental Health Perspectives 5:149-162.

Yunghans, W.N. and Morre, D.J. 1973. A rapid and reproducible homogenization procedure for the isolation of plasma membrane from rat liver. Preparative Biochemistry 3:301-312.

THE WASTING SYNDROME AND HORMONAL ALTERATIONS IN 2,3,7,8-TETRACHLORODIBENZO-*p*-DIOXIN TOXICITY

Richard E. Peterson, Ph.D.*†
Carl L. Potter, Ph.D.†
Robert W. Moore, Ph.D.*

*School of Pharmacy
and
†Environmental Toxicology Center
University of Wisconsin
Madison, Wisconsin 53706

This paper closely follows the presentations by Dr. Potter and Dr. Peterson during the symposium. It is adapted in part from articles by Seefeld and Peterson (1984), and Seefeld et al. (1984) in <u>Toxicology and Applied Pharmacology</u>.

From <u>Public Health Risks of the Dioxins</u>, proceedings of a symposium held on October 19-20, 1983 at The Rockefeller University, New York City. Edited by William W. Lowrance. Published by William Kaufmann, Los Altos, California.

INTRODUCTION

2,3,7,8-Tetrachlorodibenzo-p-dioxin (TCDD) serves as the prototype for toxicity produced by a variety of halogenated aromatic hydrocarbons that include the dibenzo-p-dioxins, dibenzofurans, azo(xy)benzenes, and biphenyls (McConnell, 1980; Poland and Knutson, 1982). A characteristic of TCDD toxicity is that its LD_{50} varies over a 5,000-fold range in common laboratory species. The guinea pig is the most sensitive to TCDD-induced lethality, the hamster is one of the least sensitive species, and rats, mice, rabbits, monkeys, and chickens exhibit intermediate sensitivity (McConnell, 1980; Poland and Knutson, 1982). The biological basis for this species difference in LD_{50} is poorly understood (Gasiewicz, 1983), because the mechanism of lethality is unknown. In the case of the guinea pig and hamster, the 5,000-fold difference in lethal potency of TCDD cannot be explained by the threefold difference in whole body elimination rate of the toxicant (Olson et al., 1980; Gasiewicz and Neal, 1979).

Another salient feature of halogenated aromatic hydrocarbon toxicity is that certain effects of these agents, such as edema and histopathologic changes in a wide variety of organs, are highly species specific whereas other effects are observed in all species. The most studied of the biochemical responses to the halogenated aromatic hydrocarbons, the induction of hepatic cytochrome P-450-mediated monooxygenase activities, occurs in all species (Goldstein, 1980) as does the prolonged wasting syndrome with loss of adipose tissue and involution of lymphoid organs (McConnell, 1980). Also in all species, the 2-3 week delay period prior to death cannot be shortened by increasing the dose of TCDD. The cause of death in TCDD-treated animals is unknown.

The mechanism by which the halogenated aromatic hydrocarbons produce biochemical and histopathological changes is not understood. The most parsimonious explanation is that TCDD and related compounds bind to a cytosol protein, termed the cytosol receptor (Poland et al., 1976). The receptor--halogenated aromatic hydrocarbon complex translocates to the nucleus and mediates the ensuing gene expression (Poland et al., 1979; Goldstein, 1980). It has been postulated by

Poland and Knutson (1982) that all organs that have the receptor respond to TCDD with a limited pleiotropic response consisting of induction of enzymes related to drug metabolism, whereas organs which show histopathologic changes respond to TCDD by expressing an additional battery of genes that regulate cell involution, altered differentiation, or proliferation. These investigators also suggest that organs which have the receptor but do not display histopathologic changes might possess these additional genes but fail to express them in response to TCDD. This cellular model of halogenated aromatic hydrocarbon toxicity is supported by studies on mouse teratoma cell XB and HRS/J mouse epidermis (Knutson and Poland, 1980; Knutson and Poland, 1982).

In the present paper, we will discuss two aspects of halogenated aromatic hydrocarbon toxicity at the whole animal level: (1) the wasting syndrome, and (2) the imbalance in circulating levels of thyroid hormones and androgens. Both effects are involved in the overall response of animals to the halogenated aromatic hydrocarbons. For the wasting syndrome, we will describe the predominant features of the progressive weight loss response in TCDD-treated rats. We will then propose, as a heuristic model of the response, that TCDD lowers a set-point for regulated body weight and that hypophagia serves as a secondary response to reduce the animal's weight. We will describe the effects of halogenated aromatic hydrocarbons on the circulating levels of thyroid hormones and androgens in experimental animals and will discuss the possible role of these changes in mediating histopathologic effects of TCDD on the thyroid gland and on the male reproductive organs, respectively. In addition, we will briefly mention the effects of TCDD on circulating levels of glucocorticoids.

CHARACTERIZATION OF THE WASTING SYNDROME

The halogenated aromatic hydrocarbons cause a wasting syndrome in several animal species (McConnell, 1980). In the young the response is manifested as a cessation of weight gain whereas in the adult it is observed as weight loss. The dose of TCDD required to elicit the response varies widely among species. However, if a dose of TCDD approaching the LD_{50} is administered, death is invariably preceded by the wasting syndrome. This insidious response has been observed in laboratory animals (mouse, rat, guinea pig, rabbit, dog, and monkey), farm animals (chicken, cow, and horse), and wildlife (mink).

Recently there has been some controversy as to whether the wasting syndrome is caused entirely by hypophagia or also involves impaired absorption of nutrients from the gastrointestinal tract. Pair-feeding studies in rats have not resolved the issue because some studies report that pair-fed animals weigh significantly more than their TCDD-treated counterparts (Gasiewicz et al., 1980; Ball and Chhabra, 1981) whereas others find that the two groups of animals lose nearly the same amount of weight (Seefeld and Peterson, 1983a, 1983b; Potter et al., 1983). One possible explanation for this discrepancy is that rats treated with TCDD spill significantly more feed than control animals (Seefeld et al., 1984). Therefore, if in the earlier pair-feeding studies feed intake was not as rigorously corrected for spillage as in the more recent studies, feed intake in TCDD-treated rats might have been overestimated. This would cause the pair-fed control animals to be provided with more feed than was actually consumed by their TCDD-treated partners.

Effects of TCDD treatment on body weight, feed intake, and fecal output in rats is shown in Figure 1. Acute treatment with TCDD caused a dose-dependent and time-related decrease in body weight, feed intake, and fecal output. Thus, the anorectic effect of TCDD is definitely involved in the weight loss response.

Figure 1. Time-dependent changes in body weight, feed intake and fecal output in control and TCDD-treated rats. Male rats were treated on Day 0 with TCDD (15 or 50 μg/kg) or vehicle (0 μg/kg). (From Seefeld and Peterson, 1984, with permission from Academic Press.)

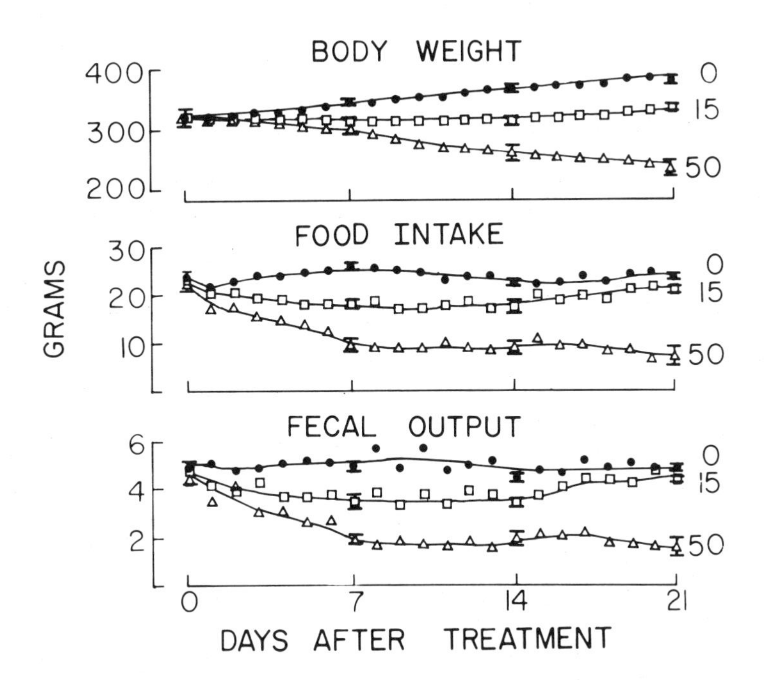

To determine if a gross malabsorption syndrome is also involved, direct calorimetry was used to determine the percentage of daily feed energy intake (kcal/day) that was eliminated in feces or absorbed by the animal as digestible energy (Figure 2). The rats used for this study were the same as in Figure 1. As expected from the previous feed intake results, daily feed energy intake was depressed by TCDD in a dose-dependent fashion. However, fecal energy loss, expressed as a percentage of daily energy intake, was similar in TCDD-treated and ad libitum-fed control rats. Likewise, digestible energy, calculated as the percentage of daily energy intake absorbed by the gut, was also similar in all groups for the duration of the study. It could be argued that these results are misleading because fecal energy loss and digestible energy should have been compared between pair-fed control rats and their TCDD-treated counterparts. That is, a pair-fed control animal might exhibit some conservative action causing a greater percentage of ingested feed energy to be absorbed. Contrary to this expectation, we found that fecal energy loss and digestible energy, expressed as a percentage of daily energy intake, in TCDD-treated (50 μg/kg) and pair-fed control rats are similar during the first 16 days post-treatment. This is the time when the majority of weight loss occurs (Potter and Peterson, unpublished results). Taken together, these findings dispel the long-standing proposal that a gross malabsorption syndrome is partially responsible for weight loss in TCDD-treated rats and place greater emphasis on hypophagia as the cause of the wasting syndrome.

Figure 2. Time-dependent changes in daily energy intake (kcal/day) and fecal energy loss and digestible energy (expressed as a percentage of daily energy intake) in control and TCDD-treated rats. Male rats were treated on Day 0 with vehicle (open bar), 15 μg/kg TCDD (stippled bar), or 50 μg/kg TCDD (solid bar). (From Seefeld and Peterson, 1984, with permission from Academic Press.)

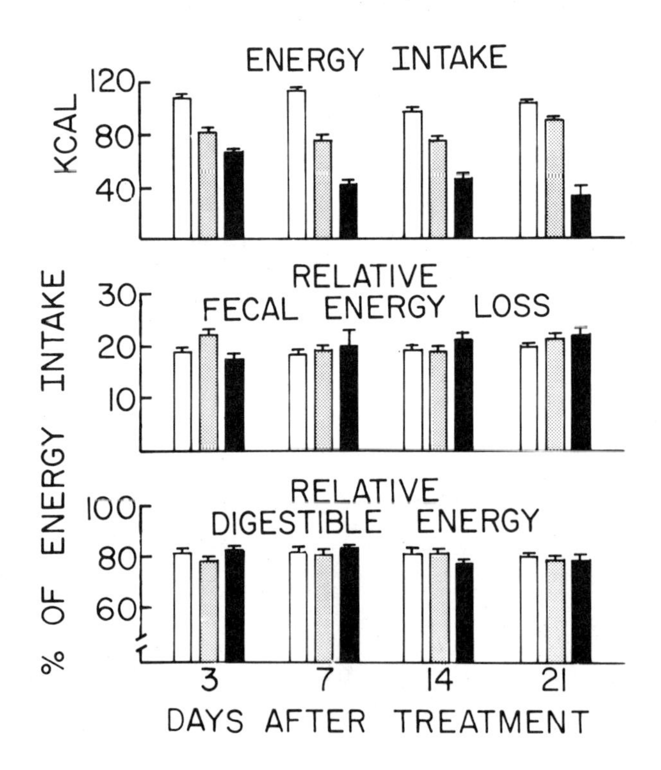

It follows from this proposed central role of hypophagia in the wasting syndrome that if the difference in energy intake between control and TCDD-treated rats is abolished, by providing both groups with the same number of calories per day by total parenteral nutrition, the weight gain in the two groups should be similar. Such a result was obtained in a total parenteral nutrition study where rats were administered a lethal dose of TCDD (Gasiewicz et al.,

1980). Thus, this finding also supports the key role played by hypophagia in TCDD-induced weight loss.

Even though a gross malabsorption syndrome does not occur in the TCDD-treated rat, we feel it is important to emphasize that our energy intake and output results (Figure 2) provide no information whatsoever on the gastrointestinal absorption of a specific dietary nutrient such as a carbohydrate, lipid, protein, vitamin or mineral. In fact, the gastrointestinal absorption of some dietary nutrients is altered by TCDD. Using in vitro intestine preparations from TCDD-treated rats, absorption of glucose was depressed (Ball and Chhabra, 1981), that of iron enhanced (Manis and Kim, 1979), and that of leucine, proline, galactose, and cobalt not affected (Schiller et al., 1982; Manis and Kim, 1977). A prolonged transit time for dietary lipid in intestinal epithelial cells of TCDD-treated rats occurs (McConnell and Shoaf, 1981) and has been associated with a decrease in intestinal protein synthesis (Shoaf and Schiller, 1981). It is postulated that the prolonged intestinal epithelial cell lipid transit is secondary to a depression in intestinal lipoprotein biosynthesis (Shoaf and Schiller, 1981), because lipoproteins are necessary for chylomicron formation and transit of lipid from intestinal epithelial cells into lymph. The significance of these various findings is that they argue for the existence of certain alterations in the intestinal absorption and/or assimilation of dietary nutrients which our energy intake-energy output study (Figure 2) was not designed to detect. These highly specific alterations in gastrointestinal function might explain why weight loss in TCDD-treated rats is not prevented when a diet of pelleted feed is supplemented by a liquid diet administered three times per day by gavage (Courtney et al., 1978), yet is completely prevented when gastrointestinal absorption is bypassed by feeding TCDD-treated rats by total parenteral nutrition (Gasiewicz et al., 1980).

Figure 3 shows that TCDD treatment reduces water intake in rats in a dose-dependent fashion without affecting urine output. These results were obtained from the same groups of animals as in Figures 1 and 2. At first, these results suggest that there is a disturbance in water balance in TCDD-treated rats because in the face of reduced water intake, urine output of the treated animals remains normal.

However, the higher than expected urine output could be a natural consequence of weight loss in the TCDD-treated rat. That is, catabolism of body tissue during weight loss causes carbon dioxide, water, and heat to be produced. Therefore, water, as a product of tissue catabolism, would be expected to be produced in greater amounts in a TCDD-treated rat that is losing weight than in a control animal that is gaining weight. Thus, the normal urine output in the TCDD-treated animal is probably a consequence of increased metabolic water production. This interpretation seems reasonable because body weight, water intake, and urine output of pair-fed control rats are nearly the same as rats treated with 50 µg/kg of TCDD (Potter and Peterson, unpublished results).

Figure 3. Water intake and urine output in control and TCDD-treated rats. Other conditions as in Figure 1. (From Seefeld and Peterson, 1984, with permission from Academic Press.)

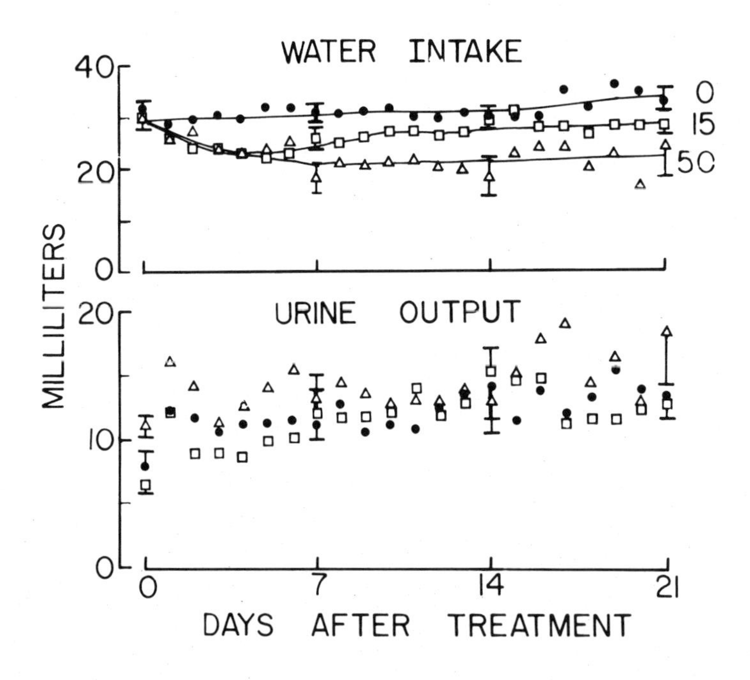

To further describe the wasting syndrome, spontaneous motor activity and oxygen consumption of TCDD-treated rats were measured. Specially designed open-circuit respirometers were used for this purpose (Figure 4). Also the animals in this particular study were evaluated at a room temperature of 27°C rather than 21°C (Figures 1, 2, and 3). The higher room temperature was selected because it is thermoneutral for rats. That is, at a room temperature of 27°C rats expend minimal energy in order to maintain body temperature at 37°C.

Figure 4. Open-circuit respirometers adapted to measure simultaneously oxygen consumption and motor activity in rats. (From Seefeld et al., 1984, with permission from Academic Press.)

Figure 5 shows, under these thermoneutral conditions, that TCDD treatment caused a dose- and time-dependent depression in body weight and feed intake (top and middle panels). Furthermore, TCDD tended to reduce the spontaneous motor activity of the animals (bottom panel).

Figure 5. Effect of TCDD treatment on body weight, feed intake, and spontaneous motor activity in rats housed at a thermoneutral room temperature. Male rats were treated on Day 0 with TCDD (15 or 50 µg/kg) or vehicle (0 µg/kg). (From Seefeld et al., 1984, with permission from Academic Press.)

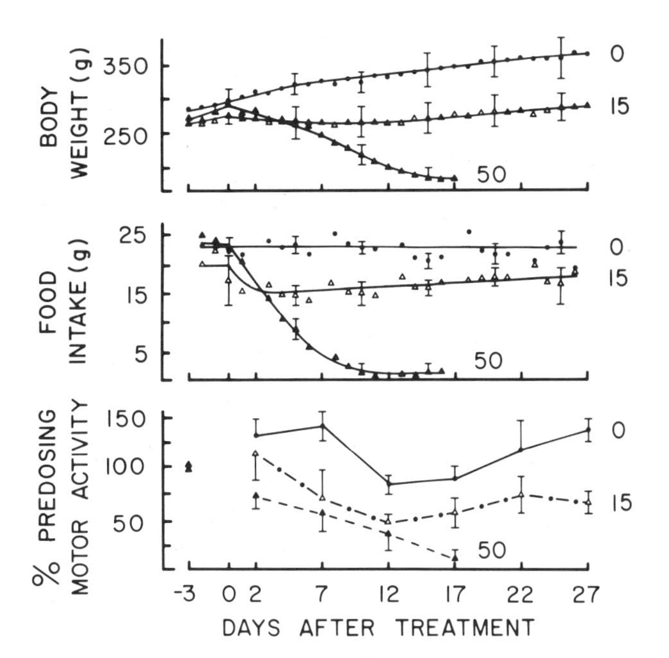

Total oxygen consumption, defined as that measured over a 24-hour period irrespective of whether the animals were resting or exhibiting motor activity, is illustrated in Figure 6. When expressed in the top panel as milliliters of oxygen consumed per minute, it can be seen that TCDD treatment caused a dose-dependent decrease in total oxygen consumption which persisted for the duration of the study. On the other hand, when corrected for differences in metabolic body size (MBS) which is defined as body weight raised to the 0.75 power (Kleiber, 1975), a different pattern of results emerged as shown in the bottom panel. Total oxygen consumption (ml O_2/MBS/min) in the low TCDD dose group eventually recovers to the same level as that of control rats while in the high TCDD dose group it remains depressed.

Figure 6. Effects of TCDD treatment on total oxygen consumption. Other conditions as in Figure 5. (From Seefeld et al., 1984, with permission from Academic Press.)

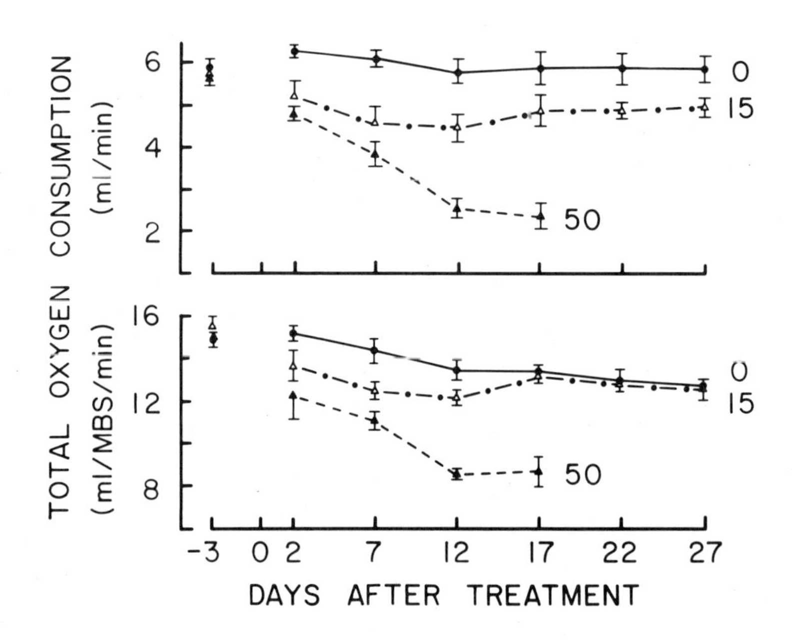

Figure 7 shows that essentially the same pattern of response to TCDD treatment is observed for resting oxygen consumption. This finding is important because it demonstrates that the lower total oxygen consumption observed for TCDD-treated rats in Figure 6 was not entirely due to their reduced level of motor activity.

Figure 7. Effects of TCDD treatment on resting oxygen consumption. Other conditions as in Figure 5. (From Seefeld et al., 1984, with permission from Academic Press).

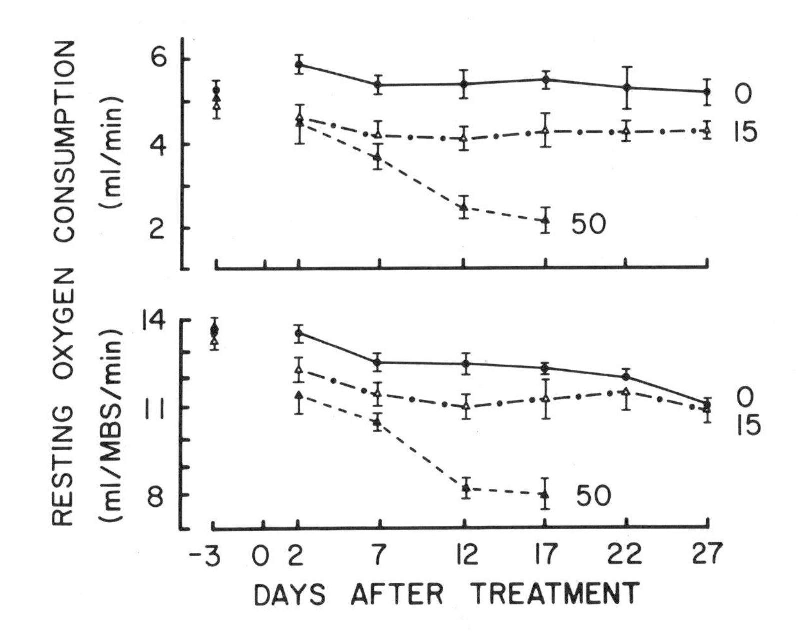

To determine if the depressant effect of TCDD treatment on oxygen consumption was secondary to its hypophagic effect, a pair-feeding experiment was conducted (results not shown). It revealed that the dose- and time-dependent nature of the depression in total and resting oxygen consumption in pair-fed control rats and rats treated with 15 or 50 μg/kg of TCDD was similar (Potter et al., 1984). Thus, the hypophagic effect of TCDD is the principal cause of the reduced oxygen consumption.

In summary, we have shown in rats that hypophagia is primarily responsible for the wasting syndrome and that a gross gastrointestinal malabsorption syndrome is not involved. TCDD-treated rats tend to have a reduced level of water intake while maintaining a normal level of urine output. The higher than expected urine output does not appear to be due to a disturbance in either water balance or renal function. Rather, it seems to be a consequence of increased metabolic water formation during weight loss in the TCDD-treated animal. TCDD treatment decreases both spontaneous motor activity and oxygen consumption, but the reduction in oxygen consumption is secondary to the hypophagic effect of TCDD.

A HEURISTIC MODEL OF THE WASTING SYNDROME

We would like to propose a hypothesis for the reduced feed intake and weight loss that occurs in TCDD-treated rats. Our proposal is that TCDD causes a dose-dependent decrease in the regulation level or set-point for body weight and that hypophagia serves, as a secondary response, to reduce the animal's weight to the lower regulation level determined by the dose of TCDD administered (Seefeld and Peterson, 1983a, 1983b; Seefeld et al., 1984; Seefeld and Peterson, 1984).

Support for this heuristic model of the wasting syndrome comes from the finding that rats treated with a sublethal dose of TCDD (15 µg/kg) chronically maintain body weight at a reduced percentage of the normal level. The subnormal level of weight maintenance is not considered to be due to a permanent stunting of the animals' growth, because control rats (pair-fed to TCDD-treated rats for more than 7 weeks) recover to near-normal levels of body weight when then permitted to feed ad libitum. Furthermore, when the reduced weight maintenance level of TCDD-treated rats (15 µg/kg) was challenged by various dietary and body weight manipulations, they displayed the same precision in defending their lower weight level as control rats displayed in defending a normal level of body weight. If challenged by changes in the caloric density or palatability of their diet, TCDD-treated and control rats displayed nearly identical adjustments in feed intake and body weight. The only difference was that these adjustments took place around a lower level of weight in the TCDD-treated animal.

Still further evidence that TCDD-treated rats regulate their weight normally, but at a reduced percentage of the body weight of control rats, derives from body composition analyses (Seefeld, Keesey, and Peterson, unpublished results). In lowering their body weight to the lower weight maintenance level, TCDD-treated rats (15 µg/kg) lose both body adipose and lean tissue. But, if induced to gain weight from this reduced level by being fed a high-calorie diet, TCDD-treated rats display the same response to the overnutrition as control rats--they add body fat. This tendency to deposit excess calories as fat, rather than to restore the previously lost lean tissue, argues that normal energy balance and body composition are achieved in the TCDD-treated rat, but only at a body weight that is abnormally low for the age and sex of the animal.

Other support for this model is that TCDD-treated rats display a normal efficiency of feed utilization but do so only at a subnormal level of body weight (Seefeld and Peterson, 1984). That is, like control rats, TCDD-treated rats increase their efficiency of feed utilization (weight gain/feed intake) when their body weight is caused to fall below the lower weight level they normally maintain.

Thus, TCDD treatment appears to lower the level of regulated body weight in the rat in a dose-dependent fashion. To achieve the lower weight level, the animal reduces feed intake. If the dose of TCDD administered is sublethal, the animal's body weight levels off at a lower, relatively constant percentage of the weight of an age- and sex-matched control rat. Thereafter, the TCDD-treated rat is postulated to regulate its lower level of weight in the same fashion that a control rat regulates its higher level of weight. If a lethal dose of TCDD is administered, we postulate that the treated rat reduces feed intake and loses body weight but dies before the lower weight regulation level, determined by the dose of TCDD administered, is reached.

Similar set-point models have been used to explain the loss of body weight during the menstrual cycle in primates (Kemnitz et al., 1982) and in rats given lateral hypothalamic lesions (Keesey et al., 1976; Keesey, 1980). Body weight set-point models have also been used to account for the dramatic loss of body weight in hibernating animals (Mrosovsky, 1974) and for the prolonged periods of anorexia that naturally occur in wild birds while incubating eggs and in wild mammals while nursing and protecting their young (Mrosovsky and Sherry, 1980).

The biochemical and physiological basis for the anorexia and regulated weight loss in TCDD-treated animals and in these various examples is not known. Elucidating the mechanism will undoubtedly be important, not only because it will advance our understanding of halogenated aromatic hydrocarbon toxicity, but, on a larger scale, it will likely provide insight into the effortless way that certain species of wildlife lose weight when feed is readily available.

EFFECTS OF TCDD ON CIRCULATING THYROID HORMONE CONCENTRATIONS

A deficiency of thyroid hormones results in decreased metabolic rate, alterations in growth and development, disturbances in water and electrolyte metabolism, altered functions of the central nervous system, skeletal muscles, and cardiovascular system, and changes in lipid metabolism (Utiger, 1979). Since some signs of TCDD toxicity in rats are common to hypothyroidism (i.e., reduced resting oxygen consumption, body temperature, and hepatic storage of vitamin A), the possibility exists for these TCDD responses to be mediated by decreased serum concentrations of thyroxine (T_4) and triiodothyronine (T_3).

It was Bastomsky (1977a) who first showed that TCDD treatment increased hepatic clearance of T_4 and decreased the circulating level of the hormone in rats. Adult male rats were treated with 25 μg/kg of TCDD and thyroid function was evaluated nine days later. Both absolute and relative weights of the thyroid gland were found to be increased, the serum concentration of T_4 was decreased, and that of thyroid stimulating hormone (TSH) was increased. The 4-hour uptake of radioactive iodine by the thyroid gland was increased, and the notion that TCDD treatment produces high uptake goiters in rats was established. In view of the effects of TCDD on T_4 and TSH, an unexpected finding in the study by Bastomsky (1977a) was that TCDD treatment significantly increased the serum concentration of T_3. We have also observed a decrease in the circulating concentration of T_4 and an increase in that of T_3 in adult male rats one week after treatment with graded single doses of TCDD ranging from 6.25 to 100 ug/kg (Potter, Moore and Peterson, unpublished results). Why this occurs is unknown. Bastomsky (1977a) suggested that the increased T_3 serum level might be secondary to increased TSH stimulation of the thyroid gland followed by increased secretion of T_3 and T_4. T_4 would then be cleared from the circulation more rapidly than T_3, leaving the T_4 serum concentration reduced and that of T_3 elevated. Alternatively, TCDD might stimulate monodeiodination of T_4 to T_3 in peripheral tissues, but an experiment to test this possibility has not been reported.

A characteristic feature of the literature on thyroid effects of TCDD and related compounds is that the decrease in serum T_4 concentration in rats is reproducible from one study to the next, but the direction of the change in serum T_3 concentration is highly variable. Both Potter et al. (1983), who measured serum thyroid hormone concentrations in young male rats one week after treatment with 45 μg/kg of TCDD, and Akoso et al. (1982), who made similar measurements in young-adult male rats exposed for thirty days to 10 ppm of 3,3',4,4',5,5'-hexabromobiphenyl (HBB), found the serum concentration of T_4 to be depressed while that of T_3 was unchanged and reduced, respectively. These effects on T_3 are not only different between the two studies; they are also different from the report by Bastomsky (1977a) which showed that circulating levels of T_3 are increased in TCDD-treated rats. The only study which has reported the effect of TCDD on the concentration of free T_4 (Clark and Horn, 1965) is that of Potter et al. (1983). They found that both the free thyroxine index and the total T_4 serum concentration were significantly depressed in young male rats one week after TCDD.

In general, high uptake goiters associated with increased serum levels of TSH, decreased serum levels of total and free T_4, and variable serum levels of T_3 have been observed in rats exposed to polychlorinated biphenyl (PCB) and polybrominated biphenyl (PBB) mixtures (Mochizuki, 1975; Bastomsky, 1977b; Collins et al., 1977; Kasza et al., 1978; Collins and Capen, 1980; Allen-Rowlands et al., 1981; Akoso et al., 1982). It has been suggested, in the case of PBB mixtures, that the major functional and morphological alterations in the thyroid gland of rats are caused by those PBB congeners that compete with TCDD for binding to its receptor (Akoso et al., 1982). Nevertheless, it is important to exercise caution in ascribing all effects of these mixtures to TCDD-like congeners, because phenobarbital-type microsomal enzyme inducers are also goitrogenic in rats (Japundzic, 1969).

While in-depth morphological studies of the thyroid gland in rats exposed to PCB and PBB have been done (Collins et al., 1977;

Kasza et al., 1978; Collins and Capen, 1980, Akoso et al., 1982), similar studies involving TCDD and related compounds are limited. Akoso et al. (1982) reported that exposure of young-adult male rats to 10 ppm of 3,3',4,4',5,5'-HBB in the diet for thirty days caused an increase in relative thyroid weight and hypertrophy and hyperplasia of thyroid follicular cells. Similar histopathological results have been reported in the previously mentioned studies in rats exposed to PCB and PBB mixtures. Since in the study by Akoso et al. (1982) the serum concentrations of both T_4 and T_3 were reduced, it was postulated that the goitrogenic activity of 3,3',4,4',5,5'-HBB was mediated in part by decreased negative feedback of thyroid hormones on the hypothalamus followed by an increase in pituitary secretion of TSH to produce the goiter.

The most commonly cited explanation for the low serum concentration of T_4 in TCDD-treated rats is that TCDD and related compounds increase the hepatic clearance of T_4 (Bastomsky, 1977a). Although the exact mechanism is not known, it is presumed that TCDD induces the UDP-glucuronyltransferase for T_4 in rat liver, because when TCDD-treated rats (25 μg/kg, 9 days posttreatment) were administered ^{131}I-labeled T_4 intravenously, a greater percentage of the ^{131}I dose was excreted in bile as the glucuronide conjugate (Bastomsky, 1977a). It is known that hepatic UDP-glucuronyltransferase for T_4 is induced by a PCB mixture in rats (Bastomsky and Murthy, 1976) but this has not been demonstrated by direct enzymatic measurement in animals treated with TCDD. Hepatic clearance of T_3 is not altered by TCDD treatment in rats (Bastomsky, 1977b). Based on studies with PCB and PBB mixtures, other possible mechanisms for the reduced serum T_4 concentration in TCDD-treated rats are that TCDD might act directly on the thyroid gland to block organification of iodine to monoiodotyrosine (Allen-Rowlands et al., 1981) or interfere with the colloid droplet-lysosome interaction that is necessary for the secretion of thyroid hormones (Collins et al., 1977; Collins and Capen, 1980).

There are certain effects of TCDD treatment in rats that various investigators have correlated with decreases in the serum concentration of T_4. Potter et al. (1983) observed in young male rats treated with 45 μg/kg of TCDD that rectal temperature was reduced 2-7 days posttreatment. This hypothermic effect of TCDD was correlated with a reduction in both the serum concentration of T_4 and the free thyroxine index. The serum concentration of T_3 was less than that of control rats but the difference was not statistically significant. Since decreases in serum levels of thyroid hormones decrease basal metabolic rate in rats, it was suggested that the decrease in rectal temperature might be secondary to the lower circulating levels of T_4 and T_3. The result of a pair-feeding experiment, which showed that the hypothermic effect of TCDD was not secondary to hypophagia (Potter et al., 1983), supported this interpretation.

Another effect of a deficiency in circulating thyroid hormones is a decrease in resting oxygen consumption. In adult male rats treated with 50 μg/kg of TCDD, there is a significant decrease in resting oxygen consumption from that of ad libitum-fed control rats one week posttreatment (Seefeld et al., 1984). But, we have had difficulty ascribing this effect to hypothyroidism. First, the plasma concentration of T_3, the more biologically potent thyroid hormone, is higher in TCDD-treated rats than in ad libitum-fed controls. Second, the plasma concentration of T_3 is also higher in TCDD-treated rats (50 μg/kg) than pair-fed controls but resting oxygen consumption in the two groups is nearly identical (Potter, Moore and Peterson, unpublished results). We think the most parsimonious explanation for these findings is that the TCDD-induced depression in resting oxygen consumption is related to hypophagia and not necessarily to hypothyroidism.

The reduced storage of vitamin A as retinol in the liver of adult male rats treated with TCDD (Thunberg et al., 1979; 1980) is another effect that might be causally related to a deficiency of circulating thyroid hormones. Certain signs of TCDD toxicity are similar to those observed in vitamin A deficiency (Thunberg et al., 1980). Hypothyroid rats store very little vitamin A in the liver whereas hyperthyroid animals store more vitamin A than control rats (Johnson and Bauman, 1947). It might be argued, therefore, that a deficiency in circulating thyroid hormones is responsible for the decreased hepatic storage of vitamin A in TCDD-treated rats. Alternatively, a decrease in resting oxygen consumption in TCDD-treated rats (Seefeld et al., 1984) might be involved. Moore (1957) has suggested that the presumed interaction between thyroid function and vitamin A metabolism might be explained by the influence of the thyroid gland on basal metabolic rate which, in turn, influences hepatic disposition of vitamin A. Whether the dose-dependent effect of TCDD treatment on hepatic storage of vitamin A (Thunberg et al., 1980) is mediated by a thyroid hormone deficiency, reduction in basal metabolic rate, or some other effect of TCDD remains to be established.

Neal and associates (1979) and Rozman and coworkers (1984) have studied the influence of thyroid hormones on TCDD-induced lethality in adult male mice and rats, respectively. On the tenet that a deficiency of thyroid hormones might be causally involved in TCDD-induced lethality, Neal et al. (1979) attempted to prevent the lethality caused by a LD_{100} dose of TCDD (200 μg/kg) in mice by treating the animals with daily doses of T_3 ranging from 50-200 μg/kg/day. They observed that T_3 treatment tended to increase feed intake and prolong survival but did not prevent lethality. On the basis of this result, Neal et al. (1979) concluded that there is no causal role for a hypothyroid state in TCDD-mediated toxicity. Rozman et al. (1984) studied the onset of lethality in rats that were thyroidectomized five weeks prior to treatment with a LD_{80} dose of TCDD. It was found that thyroidectomized rats that were treated with TCDD (100 μg/kg) survived longer than similarly treated rats that were given, immediately after TCDD treatment, weekly injections of T_4 (105 μg/kg). The cause of the delaying effect of thyroidectomy on TCDD-induced lethality is unknown.

Another approach that has been used to determine the possible role of thyroid hormones in TCDD toxicity is to determine if TCDD competes with T_3 and T_4 for thyroid hormone receptor sites. Beatty et al. (1976) examined the effect of TCDD treatment on the metamorphosis of Rana catesbeiana tadpoles which is a T_3- or T_4-dependent process. No TCDD-induced lethality or deleterious effect of TCDD on the ability of the tadpoles to complete metamorphosis was observed, in spite of the large doses of TCDD used (25-1000 μg/kg). It was concluded that TCDD does not interfere with the binding of T_3 and T_4 to thyroid hormone receptors in tadpoles.

Whether this is also the case for mammalian species that are more responsive to TCDD is unknown.

EFFECTS OF TCDD ON CIRCULATING ANDROGEN CONCENTRATIONS

We have conducted experiments to determine the effects of TCDD on selected aspects of the male reproductive system of rats, specifically, circulating androgen concentrations and androgen-dependent sex organ weights. First, a 12-day time-course study was conducted in which the doses of TCDD used were nonlethal (4.5 or 15 μg/kg). This was followed by a dose-response study conducted seven days posttreatment using 6.25, 12.5, 25, 50, or 100 μg TCDD/kg. While the second study included potentially lethal doses of TCDD, all observations were made at least one week before even the most severely affected rats would have begun to die. Pair-fed control rats were also used in both experiments, so that the extent to which hypophagia was responsible for any effects of TCDD could be estimated (Moore and Peterson, 1983; Moore et al., 1984).

Our experiments would have been of dubious value if precautions to avoid stressing the animals had not been taken. This is because short-term stress can increase plasma testosterone concentrations (Frankel and Ryan, 1981; Siegel et al., 1981), while chronic stress causes testosterone concentrations to decrease (Gray et al., 1978; Taché et al., 1980; Charpenet et al., 1981). In order to determine whether stress-induced artifacts had been successfully avoided, circulating corticosterone concentrations were assayed in all ad libitum-fed control rats in our experiments. Because corticosterone values from these animals were in agreement with basal values for non-stressed male rats published by numerous other laboratories (Krieger, 1974; Moberg et al., 1975; Aschoff, 1979), we could conclude that stress had not adversely affected the experimental results. The effects of TCDD treatment on plasma corticosterone concentrations will be discussed in the last section of this paper.

We found that the basal plasma concentrations of testosterone and dihydrotestosterone were decreased by TCDD treatment in a dose-related fashion. On Day 7, testosterone concentrations were decreased to as little as 10% of that of ad libitum-fed control rats by the highest dose of TCDD, while dihydrotestosterone concentrations were as little as one-quarter of that of the ad libitum-fed control group. The ED_{50} for both responses was about 15 µg TCDD/kg, a high nonlethal dose. From the time-course study, it was determined that decreased androgen concentrations can occur within two days after TCDD treatment and can persist for at least 12 days after dosing. While weight loss in TCDD-treated and pair-fed control rats was essentially identical at all doses tested, only about half the decreases seen in plasma androgen concentrations could be attributed to hypophagia.

Dose-related decreases in the weights of male sex organs were also observed seven days after dosing. Reductions of as much as 68% (seminal vesicles), 48% (ventral prostate), 18% (caput epididymides) and 16% (testes) from that of ad libitum-fed control rats were found. Reduced testis weights have previously been reported following exposure to chlorinated dibenzo-p-dioxins (Allen and Lalich, 1962; McConnell et al., 1978), but effects on other male sex organ weights had not been published. The decreases in testes and caput epididymides weights were essentially independent of hypophagia, while about half the decreases in seminal vesicle and ventral prostate weights could be attributed to hypophagia. The effects of TCDD on seminal vesicle weights were seen as early as two days after treatment. No recovery towards that of ad libitum-fed control rats was seen in the weights of these two accessory sex organs by Day 12. On Day 7, the ED_{50} for the decreases in seminal vesicle and ventral prostate weights was about 15 µg TCDD/kg.

Most or all of the effects of TCDD on the male reproductive system of rats which we observed appear to be a consequence of TCDD-induced decreases in plasma testosterone concentrations. Although TCDD decreases the activity of the enzyme which converts testosterone to dihydrotestosterone (Gustafasson and Ingelman-Sundberg, 1979), this change does not appear to cause the decrease in plasma dihydrotestosterone concentrations. This is because a functional deficit in the activity of C_{19} steroid 5a-reductase should increase the plasma concentration ratio of testosterone to dihydrotestosterone (Conte and Grumbach, 1979), but instead we found this ratio to be decreased by TCDD treatment. Thus, the decreased testosterone concentrations alone can account for the decreased dihydrotestosterone concentrations. The decreased accessory sex organ weights are undoubtedly a consequence of the reduced plasma androgen concentrations that followed TCDD treatment, because the weights of these organs in control animals are predominantly determined by plasma androgen concentrations. In fact, one classic bioassay technique for androgens is to determine seminal vesicle or ventral prostate weights following administration of a potentially androgenic compound to castrated or sexually-immature animals (Leathem, 1963; Mainwaring, 1977). The decreased testes and caput epididymides weights are the only responses which may be independent of the decreased androgen concentrations, since androgens, pituitary hormones, and other factors can all affect these organs (Albert, 1961; Steinberger, 1979). In summary, the TCDD-induced depression in plasma testosterone concentrations appears to be the primary event discovered, which can, in turn, account for most or all of the other effects shown.

The mechanism by which TCDD treatment decreases plasma testosterone concentrations remains unknown. At most, only half of the decrease can be attributed to hypophagia. In control rats, both total and partial starvation decrease plasma testosterone concentrations by decreasing plasma luteinizing hormone concentrations (Srebnik, 1970; Howland and Skinner, 1973; Howland, 1975; Campbell et al., 1977; Pirke and Spyra, 1981). We have recently obtained evidence that TCDD has no effect on plasma luteinizing hormone concentrations (Moore et al., 1984). Therefore, if hypophagia is the cause of the decreased testosterone concentrations seen in TCDD-treated rats, then it may have its effect via a mechanism different from that seen in feed-restricted control rats. Experiments from other laboratories on the effects of TCDD treatment on androgen metabolism (Hook et al., 1975; Lucier et al., 1975; Nienstedt et al., 1979; Gustafasson and Ingelman-Sundberg, 1979; Tofilon and Piper, 1982) provide little convincing evidence for a decreased testosterone biosynthetic capacity or an increased testosterone degradative and/or excretory capacity in male rats as being responsible for the reduction in plasma androgen concentrations.

It is important to mention that in our studies we have made no measurements of male reproductive function per se. Nevertheless, we would anticipate that the decreased androgen concentrations observed would cause male reproductive dysfunction. In work from other laboratories, a dose of TCDD which caused no body weight loss or mortality had no adverse effects on male reproductive function in rats (Murray et al., 1979). Doses which caused some mortality decreased the pregnancy rate in control female rats mated with TCDD-treated males (Khera and Ruddick, 1973). In male mice, few effects were seen at toxic but nonlethal doses of TCDD mixed with chlorinated phenoxy acids (Lamb et al., 1981 a,b,c). Our results are consistent with the results of these studies in that effects in male rodents are seen only when overt toxicity, as measured by decreased body weight gain, is observed.

Since these studies were all done using sexually-mature male rats or mice, it would be interesting to learn the effects and determine the potency of TCDD if it were given to male animals before or during the time when their reproductive system was developing. Such a study, using a polychlorinated biphenyl mixture, found that male rats were quite sensitive to polychlorinated biphenyls if exposure was begun neonatally (Sager, 1983). It would also be interesting to determine whether pure individual halogenated aromatic hydrocarbons, other than TCDD, can decrease plasma androgen concentrations and accessory sex organ weights. We predict that any such compound which competes with TCDD for binding to its cytosolic receptor (Poland and Knutson, 1982) will have similar effects on the male reproductive system.

EFFECTS OF TCDD ON CIRCULATING GLUCOCORTICOID CONCENTRATIONS

Corticosterone concentrations were also determined in the serum or plasma obtained from the rats used in our male reproductive studies (see above). As previously mentioned, basal concentrations of corticosterone were found in the ad libitum-fed control rats used in these experiments, indicating the absence of stress-induced artifacts. When assayed at the diurnal maximum, seven days after TCDD treatment, plasma corticosterone concentrations were unchanged or slightly decreased from that of ad libitum-fed control rats regardless of how much TCDD had been given. However, when assayed at the diurnal maximum on Day 12, TCDD treatment, at a high nonlethal dose, caused serum corticosterone concentrations to decrease. When assayed at the time of the diurnal minimum, nonlethal doses of TCDD were found to have no detectable effect through Day 12 (Moore and Peterson, 1983).

The results of earlier experiments to determine the effects of TCDD on glucocorticoids are described below. Vos et al. (1973) found that TCDD treatment had no effect on serum cortisol or corticosterone concentrations in guinea pigs. Neal et al. (1979) reported that an LD_{50} dose of TCDD first slightly decreased, then substantially elevated plasma corticosterone concentrations in rats 7 and 14 days after treatment. Unfortunately, neither report states what lighting schedule was used, what time of day blood was taken, or what steps were taken to avoid stressing the animals. In the absence of such information, it is difficult to draw conclusions about the biological significance of these findings.

Evidence against a major role for adrenal hyperfunction in TCDD toxicity comes from several sources. Vos and Moore (1974) found no histological evidence for adrenal cortical hyperactivity in rats, and they found that TCDD treatment decreased spleen cell graft-versus-host activity, which elevated plasma glucocorticoid concentrations would have increased. Increased plasma glucocorticoid concentrations can cause thymic atrophy, but Van Logten et al. (1980) found that TCDD-induced thymic atrophy in rats was not prevented by adrenalectomy or by hypophysectomy. In an earlier experiment, Neal et al. (1979) reported that adrenalectomy had no protective effect against TCDD-induced lethality. In fact, the opposite result was obtained. Adrenalectomy increased the lethal potency of TCDD in rats and caused the characteristic latency period between TCDD administration and death to be shortened. We have observed similar effects of adrenalectomy in TCDD-treated rats (Theobald, Moore and Peterson, unpublished results).

Glucocorticoids do not compete for binding to the TCDD receptor (Poland et al., 1976), and TCDD does not compete for the rat liver cytosol glucocorticoid receptor (Neal et al., 1979).

In summary, while we have recently obtained evidence that TCDD affects the regulation of adrenal steroidogenesis in rats (DiBartolomeis et al., 1984), no solid evidence that TCDD dramatically alters plasma total glucocorticoid concentrations has yet been reported. This finding contrasts with the profound effects of TCDD on the circulating levels of thyroid hormones and androgens in rats.

ACKNOWLEDGMENTS

This study was supported in part by National Institute of Environmental Health Sciences Grant R01-ES-01332 and by the University of Wisconsin Sea Grant College Program. R.E.P. is a recipient of a Research Career Development Award K04-ES-00098 from the National Institute of Environmental Health Sciences.

REFERENCES

Akoso, B.T., Sleight, S.D., Nachreiner, R.F. and Aust, S.D. 1982. Effects of purified polybrominated biphenyl congeners on the thyroid and pituitary glands in rats. Journal of the American College of Toxicology 1, 23-36.

Albert, A. 1961. The mammalian testis. In Sex and Internal Secretions, Third Edition (W.C. Young, ed.), pp. 305-365, Waverly Press, Baltimore.

Allen, J.R. and Lalich, J.J. 1962. Response of chickens to prolonged feeding of crude "toxic fat." Proceedings of the Society for Experimental Biology and Medicine 109, 48-51.

Allen-Rowlands, C.F., Castracane, V.D., Hamilton, M.G. and Seifter, J. 1981. Effects of polybrominated biphenyls (PBB) on the pituitary-thyroid axis of the rat. Proceedings of the Society for Experimental Biology and Medicine 166, 506-514.

Aschoff, J. 1979. Circadian rhythms: general features and endocrinological aspects. In Endocrine Rhythms (D.T. Krieger, ed.), pp. 1-61, Raven Press, New York.

Ball, L.M. and Chhabra, R.S. 1981. Intestinal absorption of nutrients in rats treated with 2,3,7,8-tetrachlorodibenzo-p-dioxin TCDD. Journal of Toxicology and Environmental Health 8, 629-638.

Bastomsky, C.H. 1977a. Enhanced thyroxine metabolism and high uptake goiters in rats after a single dose of 2,3,7,8-tetrachlorodibenzo-p-dioxin. Endocrinology 101, 292-296.

Bastomsky, C.H. 1977b. Goitres in rats fed polychlorinated biphenyls. Canadian Journal of Physiology and Pharmacology 55, 288-292.

Bastomsky, C.H. and Murthy, P.V.N. 1976. Enhanced *in vitro* hepatic glucuronidation of thyroxine in rats following cutaneous application or ingestion of polychlorinated biphenyls. *Canadian Journal of Physiology and Pharmacology 54*, 23-26.

Beatty, P.W., Holscher, M.A. and Neal, R.A. 1976. Toxicology of 2,3,7,8-tetrachlorodibenzo-p-dioxin in larvae and adult forms of *Rana catesbeiana*. *Bulletin of Environmental Contamination and Toxicology 5*, 578-581.

Campbell, G.A., Kurcz, M., Marshall, S. and Meites, J. 1977. Effects of starvation in rats on serum levels of follicle stimulating hormone, luteinizing hormone, thyrotropin, growth hormone and prolactin; response to LH-releasing hormone and thyrotropin-releasing hormone. *Endocrinology 100*, 580-587.

Charpenet, G., Taché, Y., Forest, M.G., Haour, F., Saez, J.M., Bernier, M., Ducharme, J.R. and Collu, R. 1981. Effects of chronic intermittent immobilization stress on rat testicular androgenic function. *Endocrinology 109*, 1254-1258.

Clark, F. and Horn, D.B. 1965. Assessment of thyroid function by the combined use of the serum protein bound iodine and resin uptake of ^{131}I-triiodothyronine. *Journal of Clinical Endocrinology and Metabolism 25*, 39-45.

Collins, W.T. and Capen, C.C. 1980. Fine structural lesions and hormonal alterations in thyroid glands of perinatal rats exposed *in utero* and by the milk to polychlorinated biphenyls. *American Journal of Pathology 99*, 125-142.

Collins, W.T., Capen, C.C., Kasza, L., Carter, C. and Dailey, R. 1977. Effect of polychlorinated biphenyl (PCB) on the thyroid gland of rats. *American Journal of Pathology 89*, 119-136.

Conte, F.A. and Grumbach, M.M. 1979. Pathogenesis, classification, diagnosis, and treatment of anomalies of sex. In *Endocrinology* (L.J. DeGroot, G.F. Cahill, Jr., L. Martini, D.H. Nelson, W.D. Odell, J.T. Potts, Jr., E. Steinberger, and

A.I. Winegrad, eds.), vol. 3, pp. 1317-1351, Grune and Stratton, New York.

Courtney, K.D., Putnam, J.P. and Andrews, J.E. 1978. Metabolic studies with TCDD (dioxin) treated rats. Archives of Environmental Contamination and Toxicology 7, 383-396.

DiBartolomeis, M.J., Jefcoate, C.R., Christian, B.J., Moore, R.W. and Peterson, R.E. 1984. The effect of 2,3,7,8-tetrachlorodibenzo-p-dioxin (TCDD) on adrenal steroidogenesis, in vivo. The Toxicologist 4, 187.

Frankel, A.I. and Ryan, E.L. 1981. Testicular innervation is necessary for the response of plasma testosterone levels to acute stress. Biology of Reproduction 24, 491-495.

Gasiewicz T.A. 1983. Receptors for 2,3,7,8-tetrachlorodibenzo-p-dioxin: Their inter- and intra- species distribution and relationship to the toxicity of this compound. In Proceedings of the Thirteenth Annual Conference on Environmental Toxicology, AFAMRL-TR-82-101, pp. 250-269, Air Force Aerospace Medical Research Laboratory, Wright-Patterson AFB, Ohio.

Gasiewicz, T.A. and Neal, R.A. 1979. 2,3,7,8-Tetrachlorodibenzo-p-dioxin tissue distribution, excretion, and effects on clinical chemical parameters in guinea pigs. Toxicology and Applied Pharmacology 51, 329-339.

Gasiewicz, T.A., Holscher, M.A. and Neal, R.A. 1980. The effect of total parenteral nutrition on the toxicity of 2,3,7,8-tetrachlorodibenzo-p-dioxin in the rat. Toxicology and Applied Pharmacology 54, 469-488.

Goldstein, J.A. 1980. Structure-activity relationships for the biochemical effects and the relationship to toxicity. In Halogenated Biphenyls, Terphenyls, Naphthalenes, Dibenzodioxins, and Related Products (R. Kimbrough, ed.), pp. 151-190, Elsevier/North-Holland Biomedical Press, New York.

Gray, G.D., Smith, E.R., Damassa, D.A., Ehrenkranz, J.R.L. and Davidson, J.M. 1978. Neuroendocrine mechanisms mediating the suppression of circulating testosterone levels associated with chronic stress in male rats. Neuroendocrinology 25, 247-256.

Gustafsson, J.-Å. and Ingelman-Sundberg, M. 1979. Changes in steroid hormone metabolism in rat liver microsomes following administration of 2,3,7,8,-tetrachlorodibenzo-p-dioxin (TCDD). Biochemical Pharmacology 28, 497-499.

Hook, G.E.R., Orton, T.C., Moore, J.A. and Lucier, G.W. 1975. 2,3,7,8-Tetrachlorodibenzo-p-dioxin-induced changes in the hydroxylation of biphenyl by rat liver microsomes. Biochemical Pharmacology 24, 335-340.

Howland, B.E. 1975. The influence of feed restriction and subsequent refeeding on gonadotropin secretion and serum testosterone levels in male rats. Journal of Reproduction and Fertility 44, 429-436.

Howland, B.E. and Skinner, K.R. 1973. Effect of starvation on gonadotropin secretion in intact and castrated male rats. Canadian Journal of Physiology and Pharmacology 51, 759-762.

Japundzic, M.M. 1969. The goitrogenic effect of phenobarbital-Na on the rat thyroid. Acta Anatomica 74, 88-96.

Johnson, R.M. and Bauman, C.A. 1947. Effect of thyroid on conversion of carotene into vitamin A. Journal of Biological Chemistry 171, 513-521.

Kasza, L., Collins, W.T., Jr., Capen, C.C., Garthoff, L.H. and Friedman, L. 1978. Comparative toxicity of polychlorinated biphenyl and polybrominated biphenyl in the rat thyroid gland: Light and electron microscopic alterations after subacute dietary exposure. Journal of Environmental Pathology and Toxicology 1, 587-599.

Keesey, R.E. 1980. A set-point analysis of the regulation of body weight. In Obesity (A.J. Stunkard, ed.), pp. 114-165, Saunders, Philadelphia.

Keesey, R.E., Boyle, P.C., Kemnitz, J.W. and Mitchel, J.S. 1976. The role of the lateral hypothalamus in determining the body weight set-point. In Hunger: Basic Mechanisms and Clinical Implications (D. Novin, W. Wyrwicka and G. Bray, eds.), pp. 243-255, Raven Press, New York.

Kemnitz, J.W., Lindsay, K.A. and Gibber, J.R. 1982. Ovarian influences on food intake, body weight, and sucrose consumption of rhesus monkeys. International Journal of Primatology 3, 303.

Khera, K.S. and Ruddick, J.A. 1973. Polychlorodibenzo-p-dioxins: perinatal effects and the dominant lethal test in Wistar rats. In Chlorodioxins - Origin and Fate (E.H. Blair, ed.), pp. 70-84, American Chemical Society, Washington, D.C.

Kleiber, M. 1975. The Fire of Life, R.E. Krieger Co., New York.

Knutson, J.C. and Poland, A. 1980. Keratinization of mouse teratoma cell line XB produced by 2,3,7,8-tetrachlorodibenzo-p-dioxin: An in vitro model of toxicity. Cell 22, 27-36.

Knutson, J.C. and Poland, A. 1982. Response of murine epidermis to 2,3,7,8-tetrachlorodibenzo-p-dioxin: Interaction of the Ah and hr loci. Cell 30, 225-234.

Krieger, D.T. 1974. Food and water restriction shifts corticosterone, temperature, activity and brain amine periodicity. Endocrinology 95, 1195-1201.

Lamb, J.C., IV, Marks, T.A., McConnell, E.E., Abeywickrama, K. and Moore, J.A. 1981a. Toxicity of chlorinated phenoxy acids in combination with 2,3,7,8-tetrachlorodibenzo-p-dioxin in C57BL/6 male mice. Journal of Toxicology and Environmental Health 8, 815-824.

Lamb, J.C., IV, Marks, T.A., Gladen, B.C., Allen, J.W. and Moore, J.A. 1981b. Male fertility, sister chromatid exchange, and germ cell toxicity following exposure to mixtures of chlorinated phenoxy acetic acids containing 2,3,7,8-tetrachlorodibenzo-p-dioxin. Journal of Toxicology and Environmental Health 8, 825-834.

Lamb, J.C., IV, Moore, J.A., Marks, T.A. and Haseman, J.K. 1981c. Development and viability of offspring of male mice treated with chlorinated phenoxy acids and 2,3,7,8-tetrachlorodibenzo-p-dioxin. Journal of Toxicology and Environmental Health 8, 835-844.

Leathem, J.H. 1963. Nutrition and accessory sex glands. National Cancer Institute Monograph 12, 201-210.

Lucier, G.W., McDaniel, O.S. and Hook, G.E.R. 1975. Nature of the enhancement of hepatic uridine diphosphate glucuronyl-transferase activity by 2,3,7,8-tetrachlorodibenzo-p-dioxin in rats. Biochemical Pharmacology 24, 325-334.

Mainwaring, W.I.P. 1977. The Mechanism of Action of Androgens, Springer-Verlag, New York.

Manis, J. and Kim, G. 1977. Induction of intestinal iron transport by 2,3,7,8-tetrachlorodibenzo-p-dioxin, an environmental pollutant and potent inducer of aryl hydrocarbon hydroxylase. Clinical Research 25, 468A.

Manis, J. and Kim, G. 1979. Induction of iron transport by a potent inducer of aryl hydrocarbon hydroxylase, 2,3,7,8-tetrachloro-dibenzo-p-dioxin. Archives of Environmental Health 34, 141-145.

McConnell, E.E. 1980. Acute and chronic toxicity, carcinogenesis, reproduction, teratogenesis and mutagenesis in animals. In Halogenated Biphenyls, Terphenyls, Naphthalenes, Dibenzodioxins and Related Products (R.D. Kimbrough, ed.), pp. 109-150, Elsevier/North-Holland Biomedical Press, New York.

McConnell, E.E., Moore, J.A., Haseman, J.K. and Harris, M.W. 1978. The comparative toxicity of chlorinated dibenzo-p-dioxins in mice and guinea pigs. Toxicology and Applied Pharmacology 44, 335-356.

McConnell, E.E. and Shoaf, C.R. 1981. Studies on the mechanism of 2,3,7,8-tetrachlorodibenzo-p-dioxin (TCDD) toxicity-lipid assimilation. I. Morphology. Pharmacologist 23, 176.

Moberg, G.P., Bellinger, L.L. and Mendel, V.E. 1975. Effect of meal feeding on daily rhythms of plasma corticosterone and growth hormone in the rat. Neuroendocrinology 19, 160-169.

Mochizuki, Y. 1975. Effect of polychlorinated biphenyls on the thyroid function of rats. Japanese Journal of Industrial Health 17, 498-499.

Moore, R.W. and Peterson, R.E. 1983. Effects of 2,3,7,8-tetrachlorodibenzo-p-dioxin on the male reproductive system in rats. Federation Proceedings 42, 355.

Moore, R.W., Potter, C.L., Robinson, J.A. and Peterson, R.E. 1984. Effects of 2,3,7,8-tetrachlorodibenzo-p-dioxin on the pituitary-testis axis in rats. The Toxicologist 4, 187.

Moore, T. 1957. Vitamin A. Elsevier Publishing Co., Amsterdam.

Mrosovsky, N. 1974. Natural and experimental hypothalamic changes in hibernators. In Recent Studies of Hypothalamic Function (K. Lederis and K.E. Cooper, eds.), pp. 251-267, Karger, Basel.

Mrosovsky, N. and Sherry, D.F. 1980. Animal anorexias. Science 207, 837-842.

Murray, F.J., Smith, F.A., Nitschke, K.D., Humiston, C.G., Kociba, R.J. and Schwetz, B.A. 1979. Three-generation reproduction study of rats given 2,3,7,8-tetrachlorodibenzo-p-dioxin (TCDD) in the diet. Toxicology and Applied Pharmacology 50, 241-252.

Neal, R.A., Beatty, P.W. and Gasiewicz, T.A. 1979. Studies on the mechanisms of toxicity of 2,3,7,8-tetrachlorodibenzo-p-dioxin (TCDD). Annals of The New York Academy of Sciences 320, 204-213.

Nienstedt, W., Parkki, M., Uotila, P. and Aitio, A. 1979. Effect of 2,3,7,8-tetrachlorodibenzo-p-dioxin on the hepatic metabolism of testosterone in the rat. Toxicology 13, 233-236.

Olson, J.R., Gasiewicz, T.A. and Neal, R.A. 1980. Tissue distribution, excretion, and metabolism of 2,3,7,8-tetrachloro-dibenzo-p-dioxin (TCDD) in the golden syrian hamster. Toxicology and Applied Pharmacology 56, 78-85.

Pirke, K.M. and Spyra, B. 1981. Influence of starvation on testosterone-luteinizing hormone feedback in the rat. Acta Endocrinologica (Copenhagen) 96, 413-421.

Poland, A., Glover, E. and Kende, A.S. 1976. Stereospecific, high affinity binding of 2,3,7,8-tetrachlorodibenzo-p-dioxin by hepatic cytosol. Evidence that the binding species is receptor for induction of aryl hydrocarbon hydroxylase. Journal of Biological Chemistry 251, 4936-4946.

Poland, A., Greenlee, W.F. and Kende, A.S. 1979. Studies on the mechanism of action of the chlorinated dibenzo-p-dioxins and related compounds. Annals of the New York Academy of Sciences, 320, 214-230.

Poland, A. and Knutson, J.C. 1982. 2,3,7,8-Tetrachlorodibenzo-p-dioxin and related halogenated aromatic hydrocarbons: Examination of the mechanism of toxicity. Annual Review of Pharmacology and Toxicology 22, 517-554.

Potter, C.L., Sipes, I.G. and Russell, D.H. 1983. Hypothyroxinemia and hypothermia in rats in response to 2,3,7,8-tetrachloro-dibenzo-p-dioxin administration. Toxicology and Applied Pharmacology 69, 89-95.

Rozman, K., Rozman, T. and Greim, H. 1984. Effect of thyroidectomy and thyroxine on 2,3,7,8-tetrachlorodibenzo-p-dioxin (TCDD) induced toxicity. Toxicology and Applied Pharmacology 72, 372-376.

Sager, D.B. 1983. Effect of postnatal exposure to polychlorinated biphenyls on adult male reproductive function. Environmental Research 31, 76-94.

Schiller, C.M., Walden, R. and Shoaf, C.R. 1982. Studies on the mechanism of 2,3,7,8-tetrachlorodibenzo-p-dioxin toxicity: Nutrient assimilation. Federation Proceedings 41, 1426.

Seefeld, M.D. and Peterson, R.E. 1983a. 2,3,7,8-Tetrachlorodibenzo-p-dioxin-induced weight loss: A proposed mechanism. In Human and Environmental Risks of Chlorinated Dioxins and Related Compounds (R.E. Tucker, A.L. Young and A. Gray, eds.), pp. 405-413, Plenum, New York.

Seefeld, M.D. and Peterson, R.E. 1983b. The starvation-like syndrome and 2,3,7,8-tetrachlorodibenzo-p-dioxin: New ideas on the mode of action at the whole animal level. In Proceedings of the Thirteenth Annual Conference on Environmental Toxicology, ARAMRL-TR-82-101, pp. 237-249, Air Force Aerospace Medical Research Laboratory, Wright-Patterson AFB, Ohio.

Seefeld, M.D. and Peterson, R.E. 1984. Digestible energy and efficiency of feed utilization in rats treated with 2,3,7,8-tetrachlorodibenzo-p-dioxin. Toxicology and Applied Pharmacology 74, (in press).

Seefeld, M.D., Corbett, S.W., Keesey, R.E., and Peterson, R.E. 1984. Characterization of the wasting syndrome in rats treated with 2,3,7,8-tetrachlorodibenzo-p-dioxin. Toxicology and Applied Pharmacology 73, 311-322.

Shoaf, C.R. and Schiller, C.M. 1981. Studies on the mechanism of 2,3,7,8-tetrachlorodibenzo-p-dioxin (TCDD) toxicity-lipid assimilation. II. Biochemistry. Pharmacologist 23, 176.

Siegel, R.A., Weidenfeld, J., Feldman, S., Conforti, N. and Chowers, I. 1981. Neural pathways mediating basal and stress-induced secretion of luteinizing hormone, follicle-stimulating hormone, and testosterone in the rat. Endocrinology 108, 2302-2307.

Srebnik, H.H. 1970. FSH and ICSH in pituitary and plasma of castrate protein-deficient rats. Biology of Reproduction 3, 96-104.

Steinberger, E. 1979. Structural consideration of the male reproductive system. In Endocrinology (L.J. DeGroot, G.F. Cahill, Jr., L. Martini, D.H. Nelson, W.D. Odell, J.T. Potts, Jr., E. Steinberger, and A.I. Winegrad, eds.), vol. 3, pp. 1501-1509, Grune and Stratton, New York.

Taché, Y., Ducharme, J.R., Charpenet, G., Haour, F., Saez, J. and Collu, R. 1980. Effect of chronic intermittent immobilization stress on hypophyso-gonadal function of rats. Acta Endocrinologica (Copenhagen) 93, 168-174.

Thunberg, T., Ahlborg, U.G. and Johnsson, H. 1979. Vitamin A (retinol) status in the rat after a single oral dose of 2,3,7,8-tetrachlorodibenzo-p-dioxin. Archives of Toxicology 42, 265-274.

Thunberg, T., Ahlborg, U.G., Hakansson, H., Krantz, C. and Monier, M. 1980. Effect of 2,3,7,8-tetrachlorodibenzo-p-dioxin on the hepatic storage of retinol in rats with different dietary supplies of vitamin A (retinol). Archives of Toxicology 45, 273-285.

Tofilon, P.J. and Piper, W.N. 1982. 2,3,7,8-Tetrachlorodibenzo-p-dioxin-mediated depression of rat testicular heme synthesis and microsomal cytochrome P-450. Biochemical Pharmacology 31, 3663-3666.

Utiger, R.D. 1979. Hypothyroidism. In Endocrinology (L.J. DeGroot, G.F. Cahill, Jr., L. Martini, D.H. Nelson, W.D. Odell, J.T. Potts, Jr., E. Steinberger, and A.I. Winegrad, eds.), vol. 1, pp. 471-488, Grune and Stratton, New York.

Van Logten, M.J., Gupta, B.N., McConnell, E.E. and Moore, J.A. 1980. Role of the endocrine system in the action of 2,3,7,8-tetrachlorodibenzo-p-dioxin (TCDD) on the thymus. Toxicology 15, 135-144.

Vos, J.G. and Moore, J.A. 1974. Suppression of cellular immunity in rats and mice by maternal treatment with 2,3,7,8-tetrachloro-dibenzo-p-dioxin. International Archives of Allergy 47, 777-794.

Vos, J.G., Moore, J.A. and Zinkl, J.G. 1973. Effects of 2,3,7,8-tetrachlorodibenzo-p-dioxin on the immune system of laboratory animals. Environmental Health Perspectives 5, 149-162.

SOME OBSERVATIONS ON YUSHO DISEASE

Masanori Kuratsune, M.D.

Kyushu University
Fukuoka, Japan

From Public Health Risks of the Dioxins, proceedings of a symposium held on October 19-20, 1983 at The Rockefeller University, New York City. Edited by William W. Lowrance. Published by William Kaufmann, Los Altos, California.

The verification of a marker for exposure to dioxins would be valuable for epidemiologic investigation of possible prolonged ill effects from these compounds. Today I would like to propose a working hypothesis in regard to establishing such a marker. Before explaining the proposal, I will mention a few important clinical features of current Yusho patients.

Since the outbreak of the Yusho incident in 1968, 15 years have passed. During this time, the polychlorinated biphenyls (PCBs) and polychlorinated dibenzofurans (PCDFs) ingested and retained in the patients' bodies definitely have decreased.

As shown in Table 1, the current PCB levels in the blood of Yusho patients are very low, hardly different from those of the average Japanese. These levels are even much lower than those of Michigan residents (1).

Table 1. PCB blood levels in Yusho patients and healthy Japanese and Americans.

	n	Year	Mean PCB blood levels (ppb)
Yusho patients[1]	48	1980	6
Normal Japanese [1]	68	1980	3
American-Controls[2]	25	1974	22
American-Frequent fish consumers[2]	58	1974	75

1 Masuda and Kuratsune (12).

2 Humphrey (1).

PCDF levels in the patients' bodies also seem to have decreased markedly as indicated by Rappe et al. (2). Kuroki and Masuda (3) determined a total of 0.7 ppb of PCDFs in the adipose tissue of one of the two patients who died nine years after the poisoning (Table 2).

Table 2. PCDFs in adipose tissues of Yusho patients, non-Yusho patients, and normal persons.

Subject	Year of death	2,3,7,8 TCDF	2,3,4,7,8 PCDF	Other PCDFs	Total (ppb)
Yusho					
1	1969	0.6	5.7	-	9.3
2	1977	nd	0.5	0.2	0.7
3	1977	nd	nd	nd	nd
2 non-Yusho patients	1977	nd	nd	nd	nd
5 normals*					0.019 ± 0.017

*Kashimoto (4).
nd = not detected.

The level of Yusho Patient 2 is much lower than the corresponding earlier level of Patient 1 (9.3 ppb) who died about one year after the poisoning. However 0.7 ppb is much higher than the normal average level of 0.019 ppb as determined by Kashimoto (4). Table 2 also shows that the PCDF isomer of highest concentration in these patients is 2,3,4,7,8-pentachlorodibenzofuran, which is also the most toxic PCDF. A similar fact has also been observed for PCDFs in the liver of Yusho patients (3,4). It is clear that some current Yusho patients still retain PCDFs including the most toxic isomers in their bodies. Although they are in small amounts these levels are much higher than those in normal persons.

With these reduced body burdens of PCBs and PCDFs, can we consider these patients now completely well? During the past 15 years, most of the patients have slowly recovered and even the most seriously poisoned patients rarely show active dermal lesions. However, some of the patients show mitigated yet characteristic signs of poisoning, such as dark pigmentation of the skin and mucosa, unusual secretion of cheese-like material from the Meibomian gland of the conjunctiva, and occasional atheroma formation.

These persistent signs must be associated with reduced yet significantly large amounts of PDCFs that remain in the tissues of Yusho patients. Fukuyama et al. (5) recently reported that they surgically removed pigmented mucosa of the gingiva from two Yusho patients eight years after poisoning by applying a rotating sandpaper cone. Blood PCB concentration was 32 ppb in one patient, while it was undetermined in the other. To their regret, this treatment was not successful because pigmentation recurred at the same site about a year later, even though there had been no recognizable pigments left after the operation. This observation seems to indicate that the pigmented mucosa and probably also skin which affect current Yusho patients are not due to old, residual pigments but are newly-produced pigments. Another sign, according to Kohno et al. (6), in about a quarter of the current patients is a hypersecretion of cheese-like material from the Meibomian gland that has been one of the most characteristic signs of Yusho.

An important fact related to the persistence of Yusho disease is that a majority of the patients show an unusual gas chromatographic pattern of PCBs remaining in their bodies. This was first indicated by Masuda et al. (7,8).

Figure 1. Gas chromatographic patterns of PCBs in Yusho patients and normal persons.

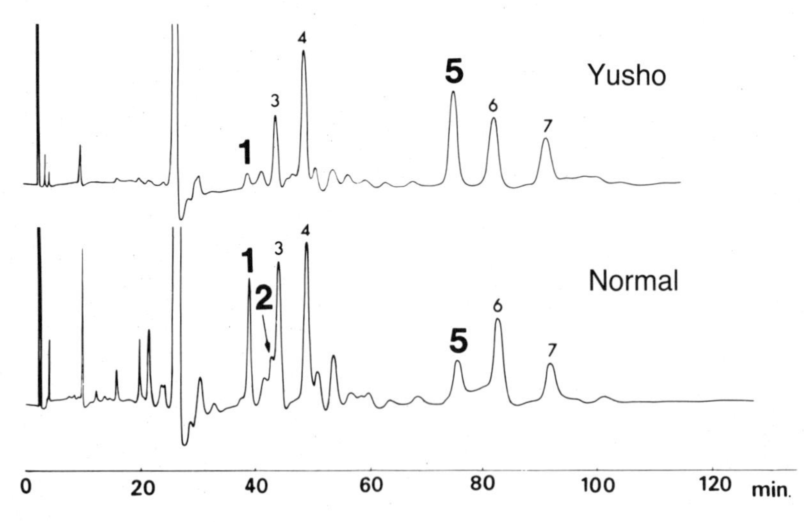

Peak 1: 2,4,5,3',4'-pentachlorobiphenyl
Peak 2: 2,3,4,3',4'-pentachlorobiphenyl
Peak 5: 2,3,4,5,3',4'-hexachlorobiphenyl

As shown in Figure 1, Peak 1 in Yusho patients is much smaller than Peak 1 in normal persons (9). Also Peak 2 almost disappears in Yusho patients but is distinct in normal persons. In contrast, Peak 5 is more prominent in Yusho patients than in normal persons. This peculiar pattern is called Pattern A for Yusho patients, while the pattern of non-Yusho persons is called Pattern C; an intermediate one is Pattern B which is seen among less severely poisoned patients (7,8,9). According to Hori et al. (10,11), the decrease of Peaks 1 and 2 is reproduced in animals by concomitant feeding of PCBs and PCDFs. This decrease is probably caused by the strong enzyme-inducing activity of PCDFs retained in the bodies of Yusho patients.

A number of studies indicate that Pattern A is closely connected with Yusho, and a person showing Pattern A can be officially recognized as Yusho even when hardly showing clinical signs and laboratory findings characteristic of the poisoning. The reasons for this are as follows:

1. Patients showing Pattern A are known to have consumed a significantly larger amount of the toxic rice oil than those showing other patterns.

2. As shown in Table 3, most of the patients showing Pattern A are more severely intoxicated.

3. Pattern A persists for a long time. As shown in Table 4, over 70% of the 49 patients showing Pattern A in 1973 still maintain Pattern A in 1980.

Table 3. PCB pattern and severity of skin lesions.

PCB Pattern	n	Severity %		
		Mild	Moderate	Severe
A	43	2.3	58.1	39.5
B	26	30.8	65.4	3.8
C	3	33.3	66.7	0.0

Table 4. 1980 pattern of 49 Yusho patients who showed Pattern A in 1973.

Pattern in 1980	n	Percent
A	36	73.5
B	12	24.5
C	1	2.0
	49	100.0

Table 5. Trend of PCB patterns in Yusho patients.

Pattern	1973-74	1982
A	43 (59.7)	40 (51.2)
B	26 (36.1)	15 (19.2)
C	3 (4.2)	23 (29.5)
Total	72 (100.0)	78 (100.0)

According to Table 5, over half of 78 Yusho patients examined in 1982 showed this same pattern A. As mentioned above, the patients have clinically recovered during the past 15 years and the majority live practically in the same way as the general population. It is conceivable, therefore, that their overall exposure to drug-metabolizing enzyme inducers must be fairly similar to that of the general population, although the exposure is without a doubt highly complex. Accordingly, Pattern A seen among Yusho patients does not seem to be explained by anything other than the PCDFs remaining in their bodies. This is particularly true if we consider the Pattern's marked persistency as well as its high prevalence among these distinct patients.

Since TCDD is an extremely strong enzyme inducer, probably stronger than PCDFs, it may alter the gas chromatographic pattern of PCBs retained in bodies exposed to it, if it persists in a significant amount in the bodies for a prolonged period of time. Alteration by TCDD exposure would result in a marked reduction of Peaks 1 and 2 just as in the case of Yusho. Thus, in those persons who are heavily exposed to TCDD, a particular gas chromatographic pattern of PCBs may be found. If this is the case, the pattern could be used as a reliable measure of past exposure to dioxins. It seems desirable, therefore, that gas chromatographic patterns of PCBs among those exposed to TCDD should be carefully examined.

A recent report indicates a substantial international variation in the relative heights of Peaks 1 to 3 of PCBs contained in human milk (13). Needless to say, the possible effects of TCDD on the height of Peaks 1 and 2 should be examined by comparing their heights relative to Peak 3, among persons exposed to TCDD, with the corresponding relative heights among the general population in each country.

Recently, I discussed this proposal with Drs. Lilis and Wolff at Mount Sinai School of Medicine. Dr. Wolff suggested that normal U.S. citizens seem to show a gas chromatographic pattern of PCBs somewhat similar to Pattern A of Yusho patients. Dr. Masuda, one of my associates, has also suggested a general tendency of Peaks 1 and 2 to decrease in the normal Japanese population, even though a distinct difference is still seen between the Yusho patients and the general Japanese. If this is true, my proposal may be wrong, but it still seems worth being tested.

REFERENCES

1. Humphrey, H.E.B. Evaluation of changes of the level of polychlorinated biphenyls (PCB) in human tissue. Michigan Department of Public Health, Lansing, Michigan, 1976.

2. Rappe, C., Nygren, M., Buser, H., Masuda, Y., Kuroki, H., and Chen, P.H. Identification of polychlorinated dioxins (PCDDs) and dibenzofurans (PCDFs) in human samples, occupational exposure and Yusho patients. Human and Environmental Risks of Chlorinated Dioxins and Related Compounds, edited by Tucker, R.E., Young, A.L., Gray, A.P. Plenum Press, New York, 241-253, 1983.

3. Kuroki, H. and Masuda, Y. Determination of polychlorinated dibenzofuran isomers retained in patients with Yusho. Chemosphere 7: 771-777, 1978.

4. Kashimoto, T. Biological effects of PCB and related compounds. Mutagens and Toxicology 14, No. 6: 24-35, 1981 (in Japanese).

5. Fukuyama, H., Anan, Y., Akamine, A., and Aono, M. Alteration in stomatological findings of patients with Yusho (PCB poisoning) in the general examination. Fukuoka Acta Medica 70: 187-198, 1979 (in Japanese).

6. Kohno, T.K., Ohnishi, Y., and Hironaka, H. Polychlorinated biphenyls in the tarsal gland contents of Yusho patients. Fukuoka Acta Medica 72: 237-241, 1981 (in Japanese).

7. Masuda, Y., Kagawa, R., and Kuratsune, M. Polychlorinated biphenyls in Yusho patients and ordinary persons. Fukuoka Acta Medica 65: 17-24, 1974 (in Japanese).

8. Masuda, Y., Kagawa, R., Shimamura, K., Takada, M., and Kuratsune, M. Polychlorinated biphenyls in the blood of Yusho patients and ordinary persons. Fukuoka Acta Medica 65: 25-27, 1974 (in Japanese).

9. Kuroki, H. and Masuda, Y. Structures and concentrations of the main components of polychlorinated biphenyls retained in patients with Yusho. Chemosphere 8: 469-474, 1977.

10. Hori, S., Miyata, H., Maeda, K., Murakami, Y., and Kashimoto, T. Studies on the characteristic patterns of PCB residues in "Yusho" patients. J. Food Hyg. Soc. Japan 20: 166-172, 1979 (in Japanese).

11. Hori, S., Kashimoto, T., and Kunita, N. Effect of polychlorinated dibenzofuran on the retention of polychlorinated biphenyl isomers in the liver and adipose tissue of mice. J. Food Hyg. Soc. Japan 23: 167-175, 1982 (in Japanese).

12. Masuda, Y. and Kuratsune, M. Unpublished data.

13. Slorach, S.A. and Vaz, R. Assessment of human exposure to selected organochlorine compounds through biological monitoring. Swedish National Food Administration, Uppsala, 1983.

THE NIOSH OCCUPATIONAL DIOXIN REGISTRY

Marilyn A. Fingerhut, Ph.D.
David A. Marlow
Patricia A. Honchar, Ph.D.
William E. Halperin, M.D., M.P.H.

National Institute for Occupational Safety and Health
Cincinnati, Ohio 45226

From Public Health Risks of the Dioxins, proceedings of a symposium held on October 19-20, 1983 at The Rockefeller University, New York City. Edited by William W. Lowrance. Published by William Kaufmann, Los Angeles, California, 1984.

The NIOSH Dioxin Registry is a compilation of demographic and work-history information for all U.S. production workers who have synthesized products known to be contaminated with 2,3,7,8-tetrachlorodibenzodioxin (2,3,7,8-TCDD) or the hexachlorinated dibenzodioxins. Currently, there are 12 production facilities and about 6,000 workers in the Registry. The first use of this information is a mortality study for which the comparison group is the U.S. male population. This study will evaluate the causes of death among workers exposed to products contaminated with dioxin.

Table 1 lists the manufactured substances relevant to the Dioxin Registry. Trichlorophenol, the herbicides 2,4,5-T and Silvex, and hexachlorophene are manufactured products which may be contaminated with 2,3,7,8-TCDD. By contrast, pentachlorophenol contains not the 2,3,7,8 isomer but the hexa-, hepta-, and octa-chlorinated dioxins. Production workers who made the herbicide 2,4-dichlorophenoxyacetic acid (2,4-D) are not included in the Registry unless they happened to be involved in one of the other processes, because 2,4-D has not been reported to contain 2,3,7,8-TCDD.

Table 1. The relevant substances.

MANUFACTURED PRODUCT	DIOXIN CONTAMINANT
Trichlorophenol 2,4,5-T (Acid, Ester, Amine) Silvex (Acid, Ester, Amine) Hexachlorophene	2,3,7,8-Tetrachlorodibenzodioxin (2,3,7,8-TCDD)
Pentachlorophenol	Hexachlorodibenzodioxins Heptachlorodibenzodioxins Octachlorodibenzodioxin
2,4-D	None Dichlorodibenzodioxins Trichlorodibenzodioxins Tetrachlorodibenzodioxins, (1,3,6,8 or 1,3,6,9)

Figure 1 illustrates the production processes in which 2,3,7,8-TCDD contamination arises. Tetrachlorobenzene is converted to the product 2,4,5-trichlorophenol (TCP), which is used as a feed stock to generate the herbicide 2,4,5-T. Under conditions of high pressure and temperature and alkalinity, the unintended dioxin contaminants are also generated. The specific isomers of dioxin produced are determined by the position of the chlorines in the reacting compounds. Both 2,4-D and pentachlorophenol are made by a different process, the chlorination of phenol, and the dioxin isomers which contaminate these products do not include the 2,3,7,8-TCDD. Hexachlorophene is made from 2,4,5-trichlorophenol by a process which does not add any additional dioxin to the amount present in the trichlorophenol. The environmental problem in Missouri has resulted from the spraying of oily dioxin wastes which were removed from trichlorophenol prior to its use in synthesizing hexachlorophene.

Figure 1. Production of TCDD from trichlorophenol.

1,2,4,5 - Tetrachlorobenzene

2,4,5 - Trichlorophenol

2,3,7,8, - Tetrachlorodibenzo-p-dioxin present as a contaminant

2,4,5 - Trichlorophenoxyacetic Acid

Hexachlorophene

2,3,7,8 - Tetrachlorodibenzo-p-dioxin present as a contaminant

2,3,7,8 - Tetrachlorodibenzo-p-dioxin present as a contaminant

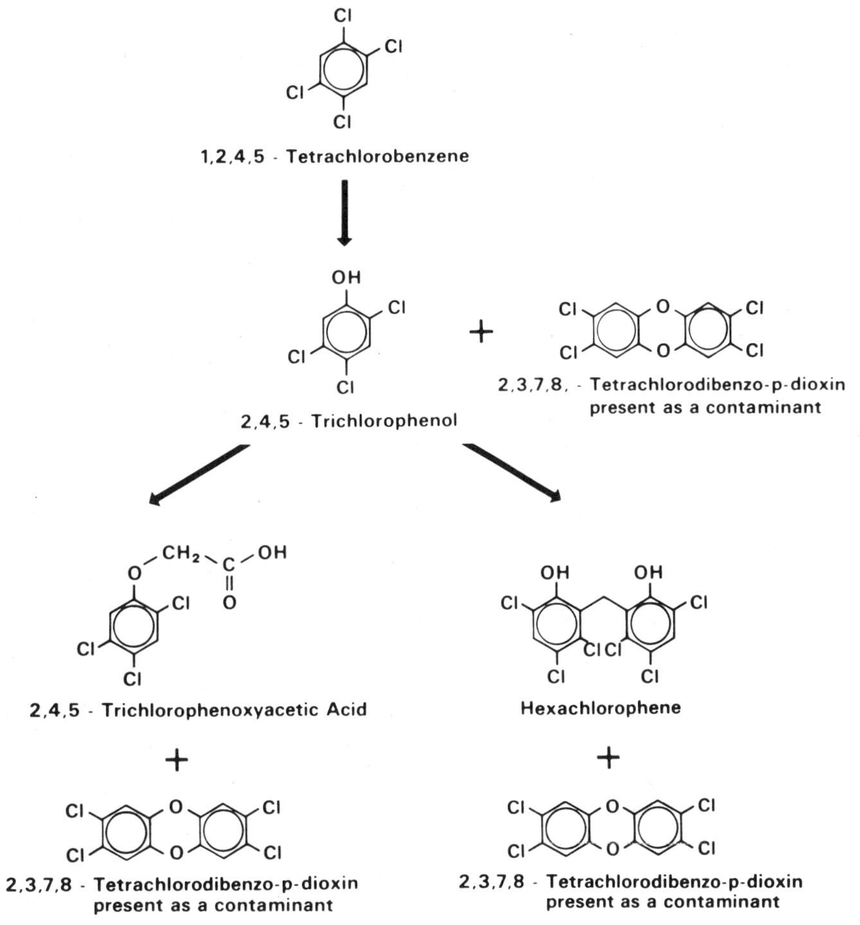

Figure 2 lists the manufacturing sites in the Registry and the years during which they produced dioxin-contaminated substances. We have identified about 6,000 production workers at these 12 chemical plants. To the best of our knowledge, this constitutes all of the manufacturing sites in the United States which synthesized the dioxin-contaminated products. There were many other places in the United States where formulation occurred; for example, where 2,4,5-T was mixed with 2,4-D or other substances for sale under a brand name. For logistical reasons, we omitted the formulation sites because most were small facilities with inadequate records. We have included formulators at the major manufacturing sites who formulated only phenoxy herbicides.

Figure 2. Years of phenoxy acid and chlorophenol production at twelve U.S. production sites.

Table 2 illustrates the types of substances produced at each site. Some companies made trichlorophenol and sold it to other companies which used the trichlorophenol to make 2,4,5-T products. We have listed 2,4-D because most of the plants which made 2,4,5-T also made 2,4-D, and many of the workers made both products in the same equipment. Pentachlorophenol was made in four facilities. We will separate the analysis of causes of death in the workers who made pentachlorophenol from the analysis of deaths among workers who made

TCP and its derivatives, because the contaminating dioxin isomers are different. We included pentachlorophenol production workers in the Dioxin Registry because there has been very little research on humans exposed to hexachlorinated dibenzodioxins.

Table 2. Summary of production with potential for dioxin contamination.

PRODUCTION SITES	TCP	T ACID	T ESTERS	T AMINES	SILVEX PRODUCTS	TCP DERIVATIVES	2,4-D PRODUCTS	PCP
01	x	x	x	x			x	
02	x	x				x		
03	x	x	x	x	x		x	
04		x	x	x	x		x	
05	x	x	x	x			x	
06	x		x	x	x		x	
07			x				x	x
08	x	x						
09	x	x	x	x	x	x	x	x
10	x							
11								x
12								x

Because the issue of exposure is so important, we have chosen a strict criterion for entrance into the Registry. The requirement is a company record of assignment to a department which made the product of interest, such as 2,4,5-T. The potential for exposure, therefore, is to products contaminated with 2,3,7,8-TCDD or the hexachlorinated dibenzodioxins, not to dioxin alone. Maintenance workers are included in the Registry if they had a record of assignment to the area where the process was located.

We will construct an exposure matrix which estimates the potential for dioxin exposure for each worker by using the following types of information: the product, the process, the operating conditions, temperatures, and solvents. In a number of trichlorophenol facilities accidents occurred during which increased amounts of dioxin were released. We have also gathered detailed information about job descriptions. The fact that an individual worked in the process does not necessarily mean that the worker was exposed to any substance. Consider, for example, that in a system involving closed pipes and kettles which require no manual loading, only leakage might be a problem. By contrast, a different potential for exposure exists at another site where the kettles are open or the worker had to shovel out some of the solid material. We also have analyses of dioxin concentrations in the products of various manufacturers.

In the mortality study we will specifically evaluate the following four carcinogenic outcomes which have been suggested in animal and human studies: soft tissue sarcoma, lymphoma, stomach cancer and liver cancer. Additionally, we will use our Life Table Analysis System to evaluate 85 other causes of death.

Our time table calls for completion of the mortality study by September, 1985.

PROPOSED FOOD AND DRUG ADMINISTRATION APPROACH TO TOLERANCE-SETTING FOR DIOXINS IN FOODS

Robert Scheuplein, Ph.D.

U.S. Food and Drug Administration
Washington, DC 20204

This paper closely follows a statement by Sanford A. Miller, Director, Bureau of Foods, FDA, presented before the Subcommittee on Natural Resources, Agriculture Research and Environment, Committee on Science and Technology, U.S. House of Representatives, June 30, 1983.

From Public Health Risks of the Dioxins, proceedings of a symposium held on October 19-20, 1983 at The Rockefeller University, New York City. Edited by William W. Lowrance. Published by William Kaufmann, Los Altos, California, 1984.

A few years ago TCDD was discovered in fish taken in and around the Great Lakes. Several of the Great Lakes States requested FDA to indicate the levels of TCDD contamination that should trigger concern. Since the Great Lakes are shared with Canada, we attempted to coordinate our concerns with those of the Canadian Health Protection Board.

We did not attempt to establish a tolerance level for TCDD in fish and we do not have tolerances today. FDA regards the contamination of these fish as a local problem, confined primarily to the Great Lakes States. Were FDA to establish a tolerance, such formal action would be challengeable in the courts and we would need to defend two propositions: (1) that fish are adulterated at or above the tolerance level with a poisonous or deleterious substance which may render it harmful to human health under ordinary conditions; and (2) that below the action or tolerance level there is no public health concern whatever.

These propositions would need to be established under a section of the Food, Drug and Cosmetic Act which deals with avoidable contaminants. To do this, TCDD in fish or in any other food would have to be regarded as an unavoidable contaminant, like PCBs, mercury in fish, or aflatoxin in corn or peanuts. This implies TCDD is not intentionally added to achieve a specific purpose and that it is impossible to eliminate completely. The Delaney Clause, which states that no substance may be approved as a food additive if it is shown to induce cancer in animal-feeding studies, does not apply to unavoidable contaminants. Instead FDA is required to consider the extent to which a substance is controllable by good manufacturing processes when it establishes a level to protect the public health. In effect, FDA must determine the balance between adverse health effects to the consumer and the needless reduction of the available food supply. This balancing is particularly difficult in the case of TCDD where so much uncertainty exists.

We recognize critics who argue that, because of the uncertainty of TCDD's health effects in humans and the knowledge of its high toxicity in laboratory animals, any level at which TCDD could be detected would be too high and therefore such food should be banned by the FDA. This approach to tolerance levels would, in fact, result in banning the interstate sale or shipment of most bottom-feeding fish in the Great Lakes and possibly other species of fish such as smelt. If the analytical methodology used for detecting TCDD improves by an order of magnitude, the likelihood is that even more species of fish would be contaminated, resulting in the closing of fisheries as well. Based on our evaluation of the available data, we felt that such a banning action was not warranted. Thus, in our advisory we stated that fish containing more than 50 ppt (parts per trillion) should not be consumed, and those containing more than 25 ppt but less than 50 ppt should not be consumed more than twice a month. We further stated that, based on fish consumption data, we saw no public health problem below 25 ppt.

Based on present knowledge, we cannot reliably state whether humans are more or less sensitive than the animal models used to explore the toxicity of TCDD. Nevertheless, in recommending the concern levels for TCDD in Great Lakes fish at 25 and 50 ppt, the FDA has in effect acknowledged the extreme toxicity of TCDD as determined experimentally in animal models.

We arrived at the level of 25 ppt by examining the distribution pattern of TCDD in Great Lakes fish to determine the risk associated with various levels of contamination. From our examination in the affected area we concluded that the species of most concern were the bottom-feeding fish such as carp and catfish. Lakes Huron, Superior, St. Clair and Michigan were sampled as well as many of the Michigan rivers including Kalamazoo, St. Joseph, Tittabawassee and Saginaw. Incidentally FDA has sampled fish from other lakes and rivers in the country and has not detected significant contamination. Looking at the levels of contamination and keeping in mind our statutory requirement to consider both public health and the availability of food, we evaluated several possible levels of concern ranging from 10 ppt to 100 ppt. We determined that 25 ppt represented a level of risk acceptable to the public health while resulting in a minimal impact on the availability of Great Lakes fish as a food source. We arrived at the 25 ppt figure in the following way.

According to food consumption surveys, the upper 90 percentile of fresh water fish consumption in the Great Lakes States is 15.7 grams of fish per day. This represents the estimated daily per capita consumption rate for the more persistent devotees of Great Lakes fish, i.e., the highest 10% consumers. In contrast, for the United States as a whole, the average total fish consumption is about 28.1 grams per day, most of which are marine fish that are not known to contain TCDD. Moreover, bottom feeders, such as carp and catfish, make up only a small fraction of all fresh water fish consumed. As far as we know this contamination of fish represents the only instance of detectable TCDD contamination of a food stuff. While our resources permit only limited monitoring of food for trace contaminants (our market basket surveys being the major source of information) no instance of detectable TCDD has been reported, even in samples containing several parts per million of PCPs (pentachlorophenols).

Using the worst-case assumption that all exposed individuals consume fish at the 90 percentile level, that all the fish consumed are fresh water fish from contaminated areas in the Great Lakes, and that these fish consist exclusively of bottom feeders such as catfish and carp having exactly 25 ppt of TCDD, we calculated a daily exposure to TCDD of 393 picograms from such fish. In order to assure that no sample exceeds a 25 ppt residue level, the average limit imposed on any given fishery area must be well below this and statistically would probably not exceed one third of that amount. Thus, if an individual consumed only bottom fish from the Great Lakes fisheries, that person would more likely consume fish with TCDD on the average of approximately 8 ppt. This then reduces the estimate of exposure to 130 picograms/day from such fish. Further, the average individual would probably consume bottom fish from uncontaminated areas as well as surface-feeding fish with nondetectable levels of TCDD. In all likelihood, no more than 10 percent of the fish in the average individual's diet from the affected states is likely to be composed of bottom-feeding fish from contaminated areas. Taking this correction into account, the estimated consumption of contaminated fish is reduced to about 13 picograms/day, or 0.8 ppt.

To determine the risks of cancer from this exposure, we utilized the animal data developed by Richard Kociba (Reference: R.J. Kociba et al., Toxicology and Applied Pharmacology 46:279-303, 1978). We regard these data as the most reliable for risk assessment, because it provides the firmest information on the relationship of dose to response. Since the mechanism by which TCDD induces cancer in rodents is unknown, it is not possible to say which one of the standard mathematical models is appropriate to use. We do assume, however, that most models are far more likely to overestimate risk rather than underestimate it. Further, risk in this context means only that the data are consistent with the probability that a risk at this level could occur, not that it necessarily will occur. It also must be kept in mind that the sensitivity of humans compared to that in animal models is not known and there is considerable controversy in scientific circles on this point.

With these caveats in mind, the lifetime risk of an 80 kg individual exposed to a 25 ppt maximum residue level is approximately three cancers out of a million (or 3×10^{-6}) based on the linear model. Given the fact that the FDA views these "concern levels" as interim statements, it is important to consider what the risk of cancer might be for less than a lifetime exposure to dioxin. For a period of one year, the risk is four cancers in a hundred million (or 4×10^{-8}) using the linear model. It should be emphasized that these calculations do not represent an actuarial risk. The intent of the calculation is to estimate the upper limit of risk that TCDD poses to exposed human beings.

With TCDD, we lack information regarding the actual sensitivity of human beings, the probable exposure to those who consume Great Lakes fish, the fraction of the population who might consume such fish and who actually do, as well as an understanding of how to extrapolate from high to extremely low doses. Only time and additional information will provide the insight necessary to adjust these estimates in whatever direction appears appropriate. More recent monitoring of TCDD levels in Great Lakes fish indicates decreasing contamination.

A final point needs to be made. At the present time, the concern level of 25 ppt is consistent with the demands and the constraints imposed by the sensitivity of the analytical method used to detect TCDD. While there are procedures which can detect TCDD at levels at approximately 1 ppt, they do not specifically identify TCDD. Confirmation that the chemical entity detected is in fact TCDD requires the use of specific mass spectroscopic techniques. In order to both detect and confirm TCDD, residue levels must be at least 10 ppt or higher. Therefore, in setting a concern level, it was necessary that our chemists who analyze fish be able to detect levels that were significantly below the concern level in order to demonstrate the distribution of fish contamination. In other words, the 25 ppt concern level for Great Lakes bottom fish requires the most sensitive method for detection and confirmation. This methodology is expensive: we estimate that it costs about $1,000 per sample. To set the concern level much lower than 25 ppt would be in fact be forcing the technology beyond its current state of the art and thus would not provide a scientific basis for legal actions that might be taken.

U.S. ENVIRONMENTAL PROTECTION AGENCY'S DIOXIN STRATEGY: EXECUTIVE SUMMARY

Overview presented by Donald G. Barnes, Ph.D.,
of the Environmental Protection Agency,
during the symposium.

The complete *Dioxin Strategy* was published by the Office of Water Regulations and Standards and the Office of Solid Waste and Emergency Response in conjunction with the Dioxin Strategy Task Force, November 28, 1983, Washington, DC 20460. Excerpted with permission of the EPA.

From *Public Health Risks of the Dioxins*, proceedings of a symposium held on October 19-20, 1983 at The Rockefeller University, New York City. Edited by William W. Lowrance. Published by William Kaufmann, Los Angeles, California, 1984.

The strategy provides a framework under which the U.S. Environmental Protection Agency (EPA) will (1) study the extent of dioxin contamination and the associated risks of humans and the environment, (2) implement or compel necessary clean-up actions at contaminated sites, and (3) further evaluate regulatory alternatives to prevent future contamination, as well as disposal atternatives to alleviate current problems.

EPA will be investigating and taking appropriate response or enforcement actions at production, disposal, and processing sites where pesticides (including herbicides) contaminated with dioxin were or are being handled. In addition, the Agency will be sampling other possibly contaminated sites as well as the ambient environment throughout the United States for the presence of dioxin. This overall investigation is in response to concerns raised by the increasing number of instances when environmental contamination by chlorinated dioxins has been documented. EPA will also continue its evaluations of human health risks associated with exposure to chlorinated dioxins and of disposal and destruction methods.

Although there are 75 different chlorinated dioxins, 2,3,7,8-tetrachlorodibenzo-p-dioxin (2,3,7,8-TCDD) is the one of primary concern because it is the most toxic dioxin isomer, with the potential of presenting significant health and disposal issues.

The 2,3,7,8-TCDD isomer is known to be a contaminant of 2,4,5-trichlorophenol (2,4,5-TCP) when 2,4,5-TCP is made from tetrachlorobenzene. 2,4,5-TCP is used in the manufacture of various phenoxy herbicides, including 2,4,5-trichlorophenoxyacetic acid (2,4,5-T) and Agent Orange, a defoliant herbicide used in Vietnam. The emphasis on 2,4,5-TCP and its derivatives is based on the fact that in nearly every place where 2,3,7,8-TCDD has been found in the environment, it can be associated, if not definitively linked, to 2,4,5-TCP production or disposal sites.

To facilitate implementation of the strategy, EPA has defined the following study tiers based on decreasing potential for 2,3,7,8-TCDD contamination:

Tier 1 - 2,4,5-TCP production sites and associated waste disposal sites.

Tier 2 - Sites (and associated waste disposal sites) where 2,4,5-TCP was used as a precursor to make pesticidal products.

Tier 3 - Sites (and associated waste disposal sites) where 2,4,5-TCP and its derivatives were formulated into pesticidal products.

Tier 4 - Combustion sources.

Tier 5 - Sites where pesticides derived from 2,4,5-TCP have been and are being used on a commercial basis.

Tier 6 - Certain organic chemical and pesticide manufacturing facilities where improper quality control on certain production processes could have resulted in the formation of 2,3,7,8-TCDD contaminated product waste streams.

Tier 7 - Control sites where contamination from 2,3,7,8-TCDD is not suspected.

The strategy calls for investigating and taking any necessary response or enforcement actions at tier 1 sites and, eventually, at tier 2 sites. Sites in tiers 3-6 will also be studied to determine the probability of contamination at these types of sites. Sampling at sites in tiers 1-6 will initially consist of a screening of areas most likely to be contaminated to determine if 2,3,7,8-TCDD is present at the site. If it is, further sampling may include all media (air, water, soil, stream sediments, fish tissue) which are appropriate to define the extent of contamination and health risk. Sampling in tier 7 will be done in two phases. In the first phase, EPA will collect multi-media samples at a number of control areas (e.g., towns, sections of cities, rural areas) selected throughout the United States. During the second phase EPA will sample fish (and other aquatic organisms) at selected stations throughout the United States. All sampling done under this strategy will follow prescribed analytical protocols.

Another important aspect of the strategy is to determine the potential health and environmental risks from exposure to 2,3,7,8,-TCDD in different media. EPA, in conjunction with other appropriate federal agencies such as the Veterans Administration (VA) and the various constituent agencies of the Department of Health and Human Services (HHS), (e.g., the Centers for Disease Control (CDC), the Food and Drug Administration (FDA), and the National Institutes for Occupational Safety and Health (NIOSH)), will undertake research to understand more fully the specific effects of 2,3,7,8-TCDD on humans and other species, and to develop techniques to determine actual risk given different levels of environmental contamination.

While investigations into the extent of human health and environmental risks from contamination by 2,3,7,8-TCDD proceed, EPA will also be evaluating different alternatives for containing and eventually disposing of soils and wastes contaminated with 2,3,7,8-TCDD. These alternatives include various methods of securing contaminated soil and preventing leachate runoff or percolation, extraction of dioxin from soils, and incineration or photolysis for destruction of dioxins.

Finally, the strategy lists a number of research activities to define the potential human health and environmental risks from dioxin isomers other than 2,3,7,8-TCDD and other "dioxin-like" chemicals. These activities include (1) assessing the toxicity of the other isomers, (2) determining their specific sources, (3) evaluating their environmental fate and transport properties, (4) developing exposure and risk assessments based on the above information, and (5) recommending appropriate control measures. To assist in these activities, the sampling program for 2,3,7,8-TCDD includes provisions to analyze for other dioxin isomers and "dioxin-like" chemicals when appropriate to the situation or the sampling site.

REGULATORY ACTIVITIES

EPA's efforts to regulate dioxin in the environment began in 1973 when the Agency instituted proceedings to cancel the registration of the pesticide 2,4,5-T, based primarily on its contamination by 2,3,7,8-TCDD. (Earlier, the U.S. Department of Agriculture had limited uses of 2,4,5-T on food crops.) EPA terminated the cancellation proceedings in 1974, partly because the analytical chemistry techniques available at the time were not capable of measuring 2,3,7,8-TCDD in food or the environment at the low levels which could pose a hazard. The Agency has since significantly improved its analytical capabilities. In 1978, EPA initiated the Rebuttable Presumption Against Registration (RPAR) process against pesticide products with 2,4,5-T. In 1979, based on a study of miscarriage rates in Alsea, Oregon (where 2,4,5-T had been sprayed on forest land) and extensive laboratory data demonstrating that 2,4,5-T, silvex, and/or 2,3,7,8-TCDD cause cancer and adverse reproductive effects in test animals, EPA ordered an emergency suspension of 2,4,5-T and silvex use on forests, rights-of-way, pastures, home gardens, turf, and aquatic vegetation. Other uses were still being evaluated under the RPAR process. Dow Chemical Company appealed the suspension in federal court and lost. In 1980, an EPA administrative law judge began consolidated cancellation hearings on the suspended and nonsuspended uses of 2,4,5-T and silvex. These hearings were postponed in 1981 to allow Dow and EPA to concentrate on settlement discussions.

Other programs have also been involved in regulatory activities related to dioxin. Under the Clean Water Act (CWA), 2,3,7,8-TCDD is listed as one of the 65 compounds and classes of compounds which EPA is required to control in industrial effluents. To date, no national discharge regulations have been issued for 2,3,7,8-TCDD, primarily because it has not been detected in effluents. The only time it has been measured in effluents was when EPA's Region 5 personnel measured it at the 50 parts per quadrillion level in the Dow effluent. (Analytical problems may be one of the reasons why 2,3,7,8-TCDD was not otherwise detected.) EPA is working with the State of Michigan on developing limitations for Dow's permit. Under the Clean Air Act, EPA is currently assessing the health impacts of 2,3,7,8-TCDD as a hazardous air pollutant.

As an interim step to control the disposal of any wastes containing 2,3,7,8-TCDD (defined as wastes resulting from the production of 2,4,5-TCP or its pesticide derivatives, or substances produced on equipment that was previously used for the production of 2,4,5-TCP or its pesticide derivatives), EPA in 1980 promulgated a rule under the Toxic Substances Control Act (TSCA) which requires any persons intending to move or otherwise dispose of these wastes to notify EPA of its plans 60 days prior to initiating any action. This allows EPA to review the plans and ensure that the wastes are properly managed. In 1983, EPA proposed to regulate wastes containing any tetra-, penta-, and hexachlorodibenzo-p-dioxins under the Resource Conservation and Recovery Act (RCRA). This action will cover a wider range of wastes and is designed to ensure that no future sites are contaminated with dioxin wastes.

As this dioxin strategy is implemented and the data are assembled, analyzed, and reviewed, various regulatory options to prevent or control future 2,3,7,8-TCDD contamination will be evaluated. Control options will include new applications of existing regulations as well as development of new regulations. Such actions as RCRA waste stream listings, CWA Section 307 (a) (2) listings, and alternative management options (e.g., prohibiting certain dioxin-containing wastes from land disposal) will be evaluated and recommendations to initiate regulatory actions will be made by appropriate program offices. Programs initiating regulatory actions should use the Dioxin Management Task Force as a steering committee for regulatory development.

MANAGEMENT AND IMPLEMENTATION OF THE STRATEGY

The Assistant Administrator (AA) for the Office of Solid Waste and Emergency Response (OSWER) under the direction of the Deputy Administrator is responsible for implementing the strategy including the periodic reporting of progress to EPA's Deputy Administrator. OSWER will directly manage the investigations and responses for sites in tiers 1 and 2. The Office of Water (OW) has been delegated responsibility for the overall management of the studies within tiers 3-7. Within tiers 3-7, individual program offices will be responsible for developing study plans relating to their programs; for example, the Office of Air, Noise and Radiation (OANR) will prepare the study plan for tier 4. The AA for OSWER will have review and approval authority for any policy or plans developed by other EPA offices that are implementing delegated portions of the strategy.

OSWER will also have oversight responsibility for the preparation of external correspondence, testimony and public statements.

This strategy reflects what is currently known about dioxins and presents a general plan for implementation. Specific work plans for the various elements of the strategy are to be prepared by the appropriate program offices in conjunction with the Regions. The actual detailed sampling plans for sites in tiers 3-7 (exclusive of tier 4) will be prepared by the Regional Offices in conjunction with the States and will be reviewed by the appropriate program office. Sampling plans for sites in tiers 1 and 2 will be prepared by the Regional Offices in conjunction with the States, CDC and NIOSH when appropriate. The individual work plans are to be more specific than the strategy and the detailed sampling plans are, by nature, unique to each sampled site; they reflect what is known at the time of their preparation and the availability of resources. As time goes on and more information is developed, the work plans and the sampling plans are expected to reflect the assimilation of new information and go through perhaps several changes. Thus, the strategy and the elements of implementation (e.g., work plans, sampling plans, disposal guidance) will evolve as new information becomes available.

OSWER has issued detailed interim guidance to the regional offices on how to proceed with investigations on the tier 1 and tier 2 sites. This guidance differentiates between the actual production sites (tiers 1 and 2) and the associated transportation, treatment, storage, and disposal sites (referred to as tiers 1A and 2A in the guidance). The basic approach is first to collect detailed information on each of the sites from EPA and State data bases and, if necessary, from site visits and employee interviews. Initially, any new field investigations (screening) will be limited to tier 1 sites; new sampling work at sites in tiers 1A, 2, and 2A will be initiated after the information being collected has been evaluated by OSWER. Where the need for a clean-up response is identified, initial efforts are to be directed at getting potentially responsible parties to take appropriate action. If prompt and appropriate clean-up is not assured by responsible parties, EPA will respond in a manner consistent with the National Contingency Plan and the Regional work plans or seek to compel response. The priority for taking either enforcement or response actions at dioxin sites will be determined by evaluating the seriousness of the problem at the site relative to the problem at all other sites whether they include dioxin or not.

Funding for investigations and response actions for sites in tiers 1 and 2 will come from the Comprehensive Environmental Response, Compensation, and Liability Act (CERCLA), while funding for studies related to tiers 3-7 will come from a special appropriation for what is referred to as "The National Dioxin Study". If contamination is detected at sites in tiers 3-7, the data will be forwarded to OSWER for further evaluation, in accordance with the interim guidance.

Implementation of the strategy will require close coordination with a number of other federal agencies, including FDA, CDC, NIOSH, and others. OSWER is responsible for ensuring that proper coordination takes place. One of the key issues requiring interagency coordination is health and environmental effects research. OSWER, in conjunction with EPA's Office of Research and Development (ORD), is responsible for developing an initial list of research needs.

Symposium Participants

William C. Agosta, PhD
Professor of Organic Chemistry
The Rockefeller University
1230 York Avenue
New York, NY 10021

Karim Ahmed, PhD
Director
Natural Resources Defense
Council
122 East 42nd Street
New York, NY 10168

Ronald Altman, MD
Assistant Commissioner
Division of Epidemiology and
Disease Control
New Jersey Department of Health
John Fitch Plaza CN360
Room 705
Trenton, NJ 08625

Karl E. Anderson, MD
Associate Professor of
Metabolism-Pharmacology
The Rockefeller University
1230 York Avenue
New York, NY 10021

Susan G. Austin, ScD
Corporate Director of
Epidemiology
Union Carbide Corporation
Section P-2586
Old Ridgebury Road
Danbury, CT 06817

Donald G. Barnes, PhD
Science Advisor to the
Assistant Administrator
Office of Pesticides and
Toxic Substances (TS-78)
U.S. Environmental Protection
Agency
401 M Street, SW
Washington, DC 20460

J. George Bekesi, MD
Professor of Medicine
and Neoplastic Disease, and
Director, Environmental
Immunological Laboratory
Mt. Sinai School of Medicine
10 East 102nd Street, Room 336
New York, NY 10029

Charles M. Benbrook, PhD
Staff Director
Subcommittee on Department
Operations Research, and
Foreign Agriculture
Committee on Agriculture
U.S. House of Representatives
Longworth House Office
Building, Room 1301
Washington, DC 20515

Brendan Birmingham, PhD
Standards Coordinator
Hazardous Contaminants
and Standards Branch
Ministry of the Environment
Province of Ontario
135 St. Clair Avenue West
Toronto, Ontario N4V 1P5
CANADA

Etcyl H. Blair, PhD
Vice President
Director, Health and
Environmental Sciences
The Dow Chemical Company
2020 Dow Center
Midland, Michigan 48640

John Blodgett
Head, Environmental Protection
Section
Environmental, Energy, and
Natural Resources Division
Congressional Research Service
Library of Congress
Washington, DC 20540

Jon Blyth
Program Officer
Charles Stewart Mott Foundation
1200 Mott Foundation Building
Flint, MI 48502

Angela Boggs, MS
Research Assistant
Department of Health Policy
and Management
Harvard School of Public Health
677 Huntington Avenue
Boston, MA 02115

Thomas A. Burke
Director
Office of Science and Research
New Jersey Department of
Environmental Protection
190 West State Street
Trenton, NJ 08625

Robert Cardell, PhD
Professor of Anatomy and
Cell Biology
College of Medicine
University of Cincinnati
Cincinnati, OH 45267

Anthony Cerami, PhD
Professor and Head
Laboratory of Medical
Biochemistry
The Rockefeller University
1230 York Avenue
New York, NY 10021

Jonathan Cole, PhD
Professor of Sociology, and
Director of the Center for
the Social Sciences
Columbia University
New York, NY 10027

Ralph R. Cook, MD, MPH
Director of Epidemiology
U.S. Medical, Health and
Environmental Sciences
1803 Building
Dow Chemical U.S.A.
Midland, MI 48640

Susan Daum, MD
Occupational Physician
16 East 96th Street
New York, NY 10028

Jack H. Dean, PhD
Head, Department of Cell Biology
Chemical Industry Institute
of Toxicology
P.O. Box 12137
Research Triangle Park,
NC 27709

Paul F. Deisler, PhD
Vice President for Health,
Safety, and Environment
Shell Chemical Company
One Shell Plaza
P.O. Box 2463
Houston, TX 77001

William Dennigan, MD
Assistant Professor of
Dermatology
University of Missouri
School of Medicine
Columbia, MO 65212

John DiGiovanni, PhD
Assistant Professor of
Biochemistry
Cancer Center
University of Texas
Smithville, TX 78957

George Eadon, PhD
Assistant Professor
Division of Environmental
Sciences
Center for Laboratories and
Research
New York State Department
of Health
Albany, NY 12201

David L. Eaton, PhD
Assistant Professor
Department of Environmental
Health, SC-34
School of Public Health
and Community Medicine
University of Washington
Seattle, WA 98195

Marilyn Fingerhut, PhD
Epidemiologist
Division of Surveillance,
Hazard Evaluations and
Field Studies
U.S. National Institute for
Occupational Safety and Health
4676 Columbia Parkway
Cincinnati, OH 45226

A. Betty Fischmann, MBBS, MRCPEd
Chief of Dermatology
Washington Veterans
Administration Medical Center
50 Irving Street, NW
Washington, DC 20422

Marvin Friedman, PhD
Professor of Biological Sciences
Hunter College
695 Park Avenue
New York, NY 10021

Perry J. Gehring, DVM, PhD
Vice President for Agricultural
Products R&D; and Director of
Health & Environmental Sciences
U.S.A.
The Dow Chemical Company
2030 Dow Center
Midland, MI 48640

John Gierthy, PhD
Research Scientist
Center for Laboratories
and Research
New York State Department
of Health
Albany, NY 12202

Michael Gochfeld, MD
Acting Chairman
Environmental and Community
Medicine Department
University of Medicine and
Dentistry of New Jersey--Rutgers
Piscataway, NJ 08854

Michael Gough, PhD
Senior Analyst
Health Program
Office of Technology Assessment
U.S. Congress
Washington, DC 20510

Donald Grant, PhD
Toxicologist
Toxicological Evaluation Division
Health and Welfare Canada
HPB Building
Tunney's Pasture
Ottawa, Ontario K1A 012
CANADA

Joseph H. Graziano, PhD
Associate Professor of
Pharmacology
Division of Pediatric Hematology-
Oncology
College of Physicians & Surgeons
of Columbia University
630 West 168th Street
New York, NY 10032

Laura Green, PhD
Research Associate
Department of Health Policy
and Management
Harvard School of Public Health
677 Huntington Avenue
Boston, MA 02115

William F. Greenlee, PhD
Scientist, Department of Cell
Biology
Chemical Industry Institute of
Toxicology
P.O. Box 12137
Research Triangle Park, NC 27709

Steven I. Hajdu, MD
Attending Pathologist and Chief,
Cytology Service
Memorial Sloan-Kettering Cancer
Center
1275 York Avenue
New York, NY 10021

William Halperin, MD, MPH
Chief, Industry Wide Studies
Branch
U.S. National Institute for
Occupational Safety and Health
4676 Columbia Parkway
Cincinnati, OH 45226

Douglas Harding
Ontario Ministry of Labor
400 University Avenue
Toronto, Ontario M7A 1T7
CANADA

Maureen Hatch, PhD, MPH
Assistant Professor of
Epidemiology
School of Public Health
Columbia University
600 West 168th Street
New York, NY 10032

Shelly Hearne, BS
Research Assistant
Natural Resources Defense
Council
122 East 42nd Street
New York, NY 10168

C.A. Heijden
Head, Laboratory of
Carcinogenicity and
Mutagenicity
National Institute of Public
Health
P.O. Box 1
3720BA Bilthoven
THE NETHERLANDS

John Hesse, MD
Acting Chief
Center for Environmental Health Sciences
Michigan Department of Public Health
P.O. Box 30035
Lansing, MI 48909

Donald Hillman, MD
Corporate Medical Director
Diamond Shamrock Corporation
717 N. Harwood
Dallas, TX 75201

Laurie Hudson
Doctoral Candidate
Department of Toxicology
Harvard School of Public Health
665 Huntington Avenue
Boston, MA 02115

Peter Kahn, PhD
Associate Professor of Biochemistry
Department of Biochemistry
328 Litman Hall
Rutgers University
New Brunswick, NJ 08903

Lawrence Kaminsky, PhD
Chief, Biochemical and Genetic Toxicology Laboratory
Center for Laboratories and Research
New York State Department of Health
Albany, NY 12202

Han Kang, DrPH
Chief
Research Section
Agent Orange Projects Office
Department of Medicine and Surgery
U.S. Veterans Administration
Washington, DC 20420

Attallah Kappas, MD
Physician-in-Chief, and Head
Laboratory of Metabolism-Pharmacology
The Rockefeller University
1230 York Avenue
New York, NY 10021

Nathan Karch, PhD
President
Karch & Associates
7713 14th Street, NW
Washington, DC 20012

Nancy K. Kim, PhD
Director, Bureau of Toxic Substances Assessment
New York State Department of Health
Empire State Plaza
Tower Building, 14th Floor
Albany, NY 12237

Renate D. Kimbrough, MD
Medical Officer
Center for Environmental Health
U.S. Centers for Disease Control
1600 Clifton Road, NE
Atlanta, GA 30333

Florence K. Kinoshita, MD
Senior Toxicologist
Medical Department
Hercules, Inc.
Hercules Plaza
Wilmington, DE 19894

Judith Klotz, DrPH
Postdoctoral Fellow
Memorial Sloan-Kettering Cancer Center; and
Science Associate
Natural Resources Defense Council
122 East 42nd Street
New York, NY 10168

Richard J. Kociba, PhD
Senior Associate Scientist
Toxicology Research Laboratory
1803 Building
Dow Chemical U.S.A.
Midland, MI 48640

Masanori Kuratsune, MD
Professor of Public Health
Faculty of Medicine
Kyushu University
1-1, 3 Chome, Maidashi
Higashi-ku, Fukuoka-shi 812
JAPAN

Robert W. Leader, DVM
Acting Director
Center for Environmental Toxicology
East Fee Hall, Room A-136
Michigan State University
East Lansing, Michigan 48824

Joshua Lederberg, PhD
President
The Rockefeller University
1230 York Avenue
New York, NY 10021

Hilton C. Lewinsohn, MD
Assistant Corporate Medical Director
Union Carbide Corporation
Section P-2590
Old Ridgebury Road
Danbury, CT 06817

Steven C. Lewis, PhD
Toxicology Associate
Medicine and Environmental Health Department
Exxon Corporation
P.O. Box 235
East Millstone, NJ 08873

Ruth Lilis, MD
Associate Professor of Occupational and Environmental Medicine
Environmental Sciences Laboratory
Mt. Sinai School of Medicine
10 East 102 Street
New York, NY 10029

William Lower, PhD
Group Leader
Environmental Trace Substances Research Center
University of Missouri
Rural Route 3
Columbia, MO 65201

William W. Lowrance, PhD
Senior Fellow and Director
Life Sciences and Public Policy Program
The Rockefeller University
1230 York Avenue
New York, NY 10021

Paul Lundy, MS
Toxic Substances Associate
Toxic Substances Program
National Audubon Society
950 Third Avenue
New York, NY 10022

Karl Mackerer, PhD
Manager of Biochemical Toxicology
Mobil Oil Corporation
P.O. Box 1029
Princeton, NJ 08540

Bonnie M. Marinelli
Senior Attorney-Advisor
Board of Veterans Appeals
U.S. Veterans Administration
Washington, D.C. 20420

Fumio Matsumura, PhD
Director
Pesticide Research Center
Michigan State University
East Lansing, MI 48824

Donald R. Mattison, MD
Medical Officer
Pregnancy Research Branch
National Institute for Child Health and Human Development
U.S. National Institutes of Health
Building 10, ACRF 86313
Bethesda, MD 20205

Donald J. McGraw, MD, MPH
Corporate Medical Director
Koppers Company, Inc.
Pittsburgh, PA 15219

Wilbur P. McNulty, MD
Head
Division of Primate Medicine
Oregon Regional Primate Research Center
505 NW 185th Avenue
Beaverton, OR 97006

Samuel Milham, Jr., MD
Head, Epidemiology Section
Population Study Unit
Washington State Department of Social and Health Services
1212 Southeast Quince
Mail Stop ET13
Olympia, WA 98405

Robert W. Miller, MD
Chief, Clinical Epidemiology
National Cancer Institute
A-521 Landow Building
National Institutes of Health
Bethesda, MD 20205

Sonya Mirsky, MS
Librarian
The Rockefeller University
1230 York Avenue
New York, NY 10021

Marion Moses, MD
500 W University Parkway
Baltimore, MD 21210

Warren Muir, PhD
Vice President
Idea Tech Associates, Inc.
One Skyline Place, Suite 1313
5205 Leesburg Pike
Falls Church, VA 22041

Devdas Mukerjee, PhD
Senior Environmental Health
Scientist
Environmental Criteria and
Assessment Office
Environmental Research Center
U.S. Environmental Protection
Agency
26 W. St. Clair Street
Cincinnati, OH 45268

Albert E. Munson, PhD
Associate Professor of
Pharmacology and Toxicology
School of Basic Sciences
Medical College of Virginia
Virginia Commonwealth University
Box 110, MCV Station
Richmond, VA 23298

F. Jay Murray, PhD
Director of Environmental Affairs
Syntex (U.S.A) Inc.
3401 Hillview Avenue
Palo Alto, CA 94304

Charles Nauman, PhD, MPH
Environmental Scientist
Office of Health and Environmental
Assessment
U.S. Environmental Protection
Agency
401 M Street, SW
Washington, DC 20460

Robert A. Neal, PhD
President
Chemical Industry Institute
of Toxicology
P.O. Box 12137
Research Triangle Park, NC 27709

Herbert Northrop, MD
Director
Department of Occupational
Medicine
Stauffer Chemical Company
Westport, CT 06881

Kathleen O'Halloran, MPH
Policy Analyst
New York City Department of
Environmental Protection
2448 Municipal Building
New York, NY 10007

Patrick O'Keefe, PhD
Director, Dioxin Analytical
Laboratory
Center for Laboratories and
Research
New York State Department of
Health
Albany, NY 12202

William Parkin, DVM, DrPH
State Epidemiologist
Division of Epidemiology and
Disease Control
New Jersey Department of Health
John Fitch Plaza CN360
Trenton, NJ 08625

Glenn Paulson, PhD
Vice President for Science
National Audubon Society
950 Third Avenue
New York, NY 10022

Dennis J. Paustenbach, PhD
Product Health Risk Specialist
Stauffer Chemical Company
Westport, CT 06880

Frederica P. Perera, DrPH
Senior Staff Scientist
Natural Resources Defense
Council
122 East 42nd Street
New York, NY 10168

Richard Peterson, PhD
Associate Professor of
Pharmacology and Toxicology
School of Pharmacy
University of Wisconsin--Madison
425 N. Charter Street
Madison, WI 53706

Christopher Portier, PhD
National Institute of
Environmental Health Sciences
U.S. National Institutes of Health
P.O. Box 12233
Research Triangle Park, NC 27709

Carl Potter, PhD
Postdoctoral Trainee
Environmental Toxicology Center
School of Pharmacy
University of Wisconsin--Madison
425 N. Charter Street
Madison, WI 53706

Christoffer Rappe, PhD
Professor and Chairman
Department of Organic Chemistry
University of Umeå
S-901 87 Umeå
SWEDEN

Arleen B. Rifkind, MD
Professor of Pharmacology
Cornell University Medical College
New York, NY 10021

Richard A. Rifkind, MD
Director, Sloan-Kettering
Division
Graduate School of Medical
Sciences
Memorial Sloan-Kettering Cancer
Center
1275 York Avenue
New York, NY 10021

Thomas Robinson, PhD
Director, Environmental Affairs
Vulcan Chemicals
P.O. Box 7689
Birmingham, AL 35253

Kenneth Roseman, MD
Director
Occupational and Environmental
Health Services
New Jersey Department of Health
John Fitch Plaza CN360
Trenton, NJ 08625

B. Sanjsder, MD
National Institute of Public
Health
P.O. Box 1
3720BA Bilthoven
THE NETHERLANDS

Shigeru Sassa, DMedSci
Associate Professor of
Metabolism-Pharmacology
The Rockefeller University
1230 York Avenue
New York, NY 10021

Arnold Schecter, MD, MPH
Professor of Preventive Medicine
Clinical Campus
Upstate Medical Center
State University of New York
Binghamton, NY 13901

Robert Scheuplein, PhD
Deputy Director for
Toxicological Sciences
U.S. Food and Drug
Administration
HFF-101
200 C Street, SW
Washington, DC 20204

Sol Schreiber
Court Appointed Special Master,
Agent Orange Cases
U.S. District Court, Eastern
District of New York
One Pennsylvania Plaza
(Suite 4915)
New York, NY 10119

Carl Schulz, Phd
Associate Scientist
Clement Associates
1515 Wilson Blvd.
Arlington, VA 22209

Barclay M. Shepard, MD
Director
Agent Orange Projects Office
(10A7)
U.S. Veterans Administration
810 Vermont Avenue, NW
Washington, DC 20420

Ellen Silbergeld, PhD
Chief Scientist
Environmental Defense Fund
1525 18th Street
New York, NY 10027

Burton H. Singer, PhD
Professor of Mathematical Statistics
Center for Social Sciences
Columbia University
420 West 118th Street
New York, NY 10027

Kenneth Skipka
Camp, Dresser & McKee (Suite 2601)
250 Broadway, Suite 2601
New York, NY 10007

Andy Smith
National Ministries
American Baptist Churches, USA
Valley Forge, PA 09482-0857

Jerry F. Stara, DVM
Director
Environmental Criteria and Assessment Office
Environmental Research Center
Environmental Protection Agency
26 West St. Clair
Cincinnati, OH 45268

Steven Stellman, PhD
Assistant Vice President for Epidemiology
American Cancer Society
4 West 35th Street
New York, NY 10001

Acos Szakolcai
Air Resources Branch
Ministry of the Environment
880 Bay Street
Toronto, Ontario M5S 1Z8
CANADA

Michael L. Taylor, PhD
Associate Professor of Pharmacology/Toxicology
Brehm Laboratory
Wright State University
Dayton, OH 45435

David E. Tenzer
Professional Staff
Media Resource Service
Scientists' Institute for Public Information
355 Lexington Avenue
New York, NY 10017

Stephen R. Thomas, PhD
Assistant Professor of Political Science and Environmental Policy
Department of Health Policy and Management
Harvard School of Public Health
677 Huntington Avenue
Boston, MA 02115

Thomas O. Tiernan, PhD
Professor of Chemistry and Director
Brehm Laboratory
Wright State University
Dayton, OH 45435

Michael Torrusio
Special Legislative Assistant on Energy and Environment
Office of U.S. Congressman Guy V. Molinari
Fort Wadsworth, Building 203
Staten Island, NY 10305

William A. Toscano, PhD
Assistant Professor of Toxicology
Harvard School of Public Health
665 Huntington Avenue
Boston, MA 02115

Helle Tosine, PhD
Advanced Instrumentation Supervisor
Pesticides Section
Laboratory Services Branch
Ministry of the Environment
P.O. Box 213
Rexdale, Ontario M9W 5L1
CANADA

Thomas Umbreit, PhD
Research Specialist
Environmental and Community Medicine Department
University of Medicine and Dentistry of New Jersey-Rutgers
Piscataway, NJ 08854

John Van Ryzin, PhD
Professor of Biostatistics
School of Public Health
Columbia University
600 West 168th Street
New York, NY 10032

J.J. Vostal, MD
Senior Medical Research Advisor
General Motors Research
Laboratory
Warren, MI 48090

I. Bernard Weinstein, MD
Director, Division of
Environmental Sciences
Columbia University College
of Physicians and Surgeons
701 West 168th Street
New York, NY 10032

James P. Whitlock, Jr., MD
Associate Professor of
Pharmacology
School of Medicine
Stanford University
Stanford, CA 94305

John Wilkinson, PhD
Technical Director
Wood Treating Products
Reichhold Chemicals Inc.
2340 Taylor Way
Tacoma, Washington 98401

Mary S. Wolff, PhD
Assistant Professor of Chemistry
Environmental Sciences Laboratory
Department of Community Medicine
Mt. Sinai School of Medicine
One Gustave L. Levy Place
New York, NY 10029

James S. Woods, PhD
Senior Research Scientist
Batelle Human Affairs
Research Centers
4000 NE 41st Street
P.O. Box C-5395
Seattle, WA 98105

Alvin L. Young, Lt. Col., USAF, PhD
Special Assistant for
Environmental Sciences
Agent Orange Projects Office
U.S. Veterans Administration
810 Vermont Avenue, NW
Washington, DC 20420